METEORITES

SECOND EDITION

SARA RUSSELL AND MONICA GRADY

PUBLISHED BY 130127 06038821 7 MUSEUM, LONDON

First published by The Natural History Museum,
Cromwell Road, London SW7 5BD
© The Natural History Museum, London 2002
ISBN 0 565 09155 7

A catalogue record for this book is available from
the British Library

Edited by Celia Coyne
Designed by Mercer Design
Reproduction and printing by Craft Print,
Singapore

DISTRIBUTION

Australia and New Zealand
CSIRO Publishing
PO Box 1139
Collingwood, Victoria 3066
Australia

UK and the rest of the world
Plymbridge Distributors Ltd.
Plymbridge House, Estover Road
Plymouth, Devon, PL6 7PY
UK

Contents

Preface

Meteorites and micrometeorites are natural objects that survive their fall to Earth from space. Almost all meteorites are from the Asteroid Belt, coming from orbits between Mars and Jupiter, although one group of (currently) 24 meteorites comes from Mars, and another, of 18 meteorites comes from the Moon.

Approximately 4560 million years ago, asteroids, Kuiper Belt objects and comets formed together with the Sun, the planets and their satellites. Since the first edition of this book, written by Robert Hutchison and Andrew Graham, many advances have been made. There has been a suggestion that meteorites from Mars can be used to search for evidence of early life on that planet. Advances in telescope imagery have allowed us to study stars in the earliest stages of their development which we can compare to the record preserved in meteorites. These and other new developments are discussed in the context of Solar System formation and evolution. Meteorites, and components within them, carry records of all stages of Solar System history and remain our only opportunity for direct study of the primitive material from which our Solar System was built.

About the authors

Sara Russell and Monica Grady lead the meteorite and micrometeorite research programme at The Natural History Museum, London. Sara obtained her degree in Natural Sciences from Cambridge University in 1988, and her Ph.D on interstellar grains in meteorites from the Open University in 1993. Following research posts at the California Institute of Technology and the Smithsonian Institution, Sara joined the NHM in 1998. Her specialist research interests are in early Solar System processes and origins. Monica Grady obtained her degree in Chemistry and Geology from Durham University in 1979, and her Ph.D on carbon in meteorites from Cambridge University in 1982. Following research posts at Cambridge and the Open University, Monica joined the NHM in 1991. Her specialist research interests are in martian meteorites and cosmic dust. Both Sara and Monica have participated in meteorite collection expeditions to Antarctica and Australia. Sara is Principal Editor of the *Meteoritical Bulletin*, and Monica is Editor of the *Catalogue of Meteorites*.

What are meteorites?

Meteorites are natural objects that survive their fall to Earth from space. Some are metallic, but most are made of stone. They are the oldest objects that we have for study. Almost all meteorites are fragments from asteroids, and were formed at the birth of the Solar System, approximately 4560 million years ago. Although the Earth, along with the other planets, was also formed at this time, on Earth none of the original material remains: over the years it has been changed and recycled through geological activity. It is only by studying meteorites that we can learn about the processes and materials that shaped the Solar System and our planet.

Meteorites are not the same as meteors or 'shooting stars'. Meteors are tiny pieces of extraterrestrial material that burn up high in the atmosphere. It is very rare for material to be recovered on Earth from a meteor. A meteoroid is a small body travelling through space that may, or may not, land on Earth as a meteorite.

Meteorites are divided into three different types on the basis of their composition: stones (composed of minerals often found in rocks on Earth), irons (made up of iron metal alloyed with nickel), and stony-irons (as their name suggests, a mixture of stone minerals and iron metal). These are further divided as described on pp. 21 onwards.

Meteorites fall, more or less, at random over the Earth's surface. There are, however, certain places where meteorites are preserved or concentrated by natural processes. The arid environment of deserts, both hot and cold, slows down the weathering that destroys meteorites, allowing their numbers to build up over time. Meteorites are generally named after a place near to where they fell or were found. Exceptions occur in the desert regions where locality names are few; in these cases, meteorites are given the name of the geographical area in which they were found, followed by a number. The largest meteorite to

BELOW LEFT The three main types of meteorite: stone, iron and stony-iron. Irons and stony-irons are the product of melting, as are stony achondrites. Chondrites have not melted since they first aggregated.

BELOW RIGHT A meteor, or shooting star, photographed during the Leonid meteor shower in November 1998.

stone

chondrites — unmelted

achondrites

iron

melted

stony-iron

have fallen and been preserved as a single object is the Hoba iron meteorite. It remains where it fell, near Grootfontein in Namibia, and is preserved as a national monument.

Entry phenomena

A meteoroid travels in space at its cosmic velocity, about 30 km per second (67,500 mph). As soon as it reaches the Earth's atmosphere, it is slowed down by friction. At the same time, friction causes the outermost surface to heat up and melt. This usually results in a very bright fireball. Only the outermost surface of the meteoroid melts and the resulting droplets of molten meteorite are carried away as the meteoroid speeds through the air. Finally, as the meteoroid is slowed down by the atmosphere, it falls under gravity to the ground. By this stage, the molten surface has cooled rapidly to a glassy

coating, or fusion crust. The presence of a fusion crust is often characteristic of meteorites. One side of the meteorite may be curved or almost conical. This is the side, or leading edge, that pointed in the direction the meteoroid travelled as it passed through the atmosphere. It is important to stress that it is only the outermost surface of the meteorite that melts: the interior remains cool and unchanged. Meteorites are almost always cold when they land. The entry speed of a meteoroid also generates a shock wave in the atmosphere, often heard as a sonic boom, or explosion.

ABOVE The conical shape points in the direction of travel during atmospheric flight. The ridges that flow away from the centre were produced by ablation during entry.

BELOW LEFT The Hoba iron meteorite in Namibia has an estimated weight of 60 tonnes, but since much has been lost through rusting, it may have weighed up to 100 tonnes when it originally fell.

RIGHT Map showing the distribution of a selection of the 50 plus rocks found in the Mbala, Uganda meteorite strewnfield. The original direction of the fall was from approximately the north.

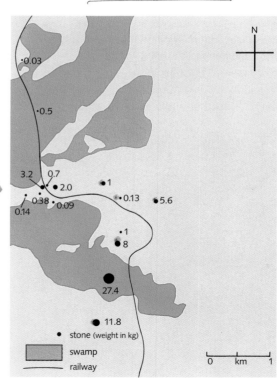

- ·0.03
- ·0.5
- 3.2 0.7
- • 2.0 •1
- 0.38 ·0.09 ·0.13 •5.6
- 0.14
- ·1
- • 8
- 27.4
- • 11.8
- • stone (weight in kg)
- ▭ swamp
- — railway

0   km   1

N

Iron meteorites are sufficiently strong that they often remain as a single body during entry. The weaker stone meteorites usually break up into smaller pieces, resulting in a shower of meteorites falling. The area over which a shower is collected is the strewnfield; larger meteorites travel farther than smaller ones, and so plotting the size distribution of meteorites within a strewnfield gives the direction of motion of the original meteoroid.

Meteorite or meteor-wrong?

Since meteorites fall almost at random over the Earth's surface, there is always a possibility that you might find one. There are, unfortunately, several objects (natural and artificial) that are frequently mistaken for meteorites (see box, right). Here is a short guide to some of the features to look for when identifying a meteorite. Although, please note that there are exceptions to every rule.

The most obvious meteorites to find are irons or stony-irons, because their appearance and nature is so different from that of terrestrial rocks. These meteorites resemble lumps of metal, are extremely dense and heavy, and are strongly attracted to magnets. Finding stony meteorites is more problematic as they can look much like terrestrial rocks. However, more often than not, they have rounded edges that have been smoothed during entry heating. But they are generally not spherical. Newly fallen or lightly weathered meteorites may have a charred fusion crust, the thin exterior coating that is usually black in colour.

Meteor-wrongs

Material	Property
Furnace slag	Quartz grains (sand); bubbles
Glass waste	Quartz grains (sand); bubbles
Natural iron oxide:	
Haematite	Non-magnetic
Magnetite	Black cubic crystals
Nodules:	
Pyrite (or marcasite)	Non-magnetic
Ironstone	Sand grains
Septarian	Non-magnetic
Tar-covered roadstone	Exterior surface soft, sticky when heated gently

ABOVE The fusion crust is a thin coating that may be glassy or matt in appearance.

The internal appearance of a suspected meteorite can also be helpful. Since weathering can wear away the fusion crust, the presence of metal is more diagnostic of meteorites. Terrestrial rocks (with very few exceptions) do not contain iron metal: the iron they contain is combined with oxygen in minerals. Practically all meteorites, including the stony ones, contain iron metal (alloyed with nickel), seen as shiny flecks inside the rock. The presence of metal draws meteorites to a magnet.

The stone might also have a rounded, droplet-looking texture on a cut surface. The droplets are called chondrules, constituents of the most common stony meteorites, but never found in terrestrial rocks.

Where are meteorites found?

There are around 20,000 meteorite falls of over 100 g every year. Most fall unobserved into the sea or into unpopulated regions; only a handful are witnessed and recovered.

Meteorites are divided into two types: 'falls' and 'finds'. Falls are meteorites that are actually observed to fall to Earth. Typically a fireball is seen, often by many people, before the meteorite lands. Meteorite finds, which are found on the ground some time after the fall took place, are often degraded by the effects of water and air. This can reduce the scientific value of a meteorite relative to a similar type seen to fall.

Meteorite falls

The rate of meteorite falls can be measured using meteor observations from around the world. Measuring the number of meteors seen around the world gives us a much better idea of the amount

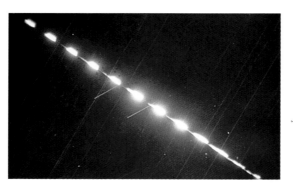

ABOVE RIGHT The Peekskill meteorite was witnessed by many people as it partially burned up in the atmosphere.

RIGHT World map showing the location of where meteorites have been seen to fall. Note that more meteorites are seen to fall in high population regions – places where there are more people to see a meteorite fall.

of extraterrestrial material coming to the Earth than the numbers of recovered meteorites, which represent only a tiny fraction of the total. The information collected shows that meteorites fall almost evenly around the whole globe, with slightly less chance of a meteorite fall near the poles of the Earth than at the equator. The lower rate at the poles results from the likelihood that extraterrestrial materials will be travelling in the plane of the planets in the Solar System. Meteorites fall at any time of the year but most falls are recorded at night, partly because they are most easily observed at this time of the day. Another possible reason is that at this point in the Earth's rotation one is facing in the same direction as the Earth is travelling around the Sun. Meteorites are therefore more likely to fall, much in the same way as flies are more likely to stick to a front windscreen when a car is moving forwards.

Although the distribution of falls around the world is probably quite even, the distribution of meteorites that are actually observed to fall is not. Most meteorites are seen to fall in places where there are people there to observe them; and so countries with a high population density tend to have the highest rates of meteorite fall recovery. Meteorites that fall into people's gardens, houses or cars are more likely to be spotted than those that fall into empty fields or forests. Countries with a strong scientific interest in meteorites (such as France and the United States) tend to have the highest rates of reported meteorite falls. The total number of recorded meteorite falls was 1014 to the end of the year 2001.

Meteorite finds

The global distribution of meteorites that are found some time after they have fallen is quite different from the distribution of meteorites that are witnessed falling. Meteorites are far more likely to be found in regions with an arid climate than in temperate areas. This is because most meteorites rust and disintegrate rather quickly in wet and humid conditions. In deserts, however, the meteorites can remain intact for thousands of years or more, and so gradually accumulate over time.

Nearly all iron meteorites so far collected have been finds, rather than falls. This is partly because iron meteorites are quite resilient to terrestrial conditions and so can survive for a long time, but also because they look completely different from terrestrial rocks, and so can be easily spotted amongst other rocks.

Systematic meteorite collection expeditions over the last three decades have found ten times more meteorites than the total of those previously known. These expeditions have focused on hot desert regions and the cold desert of Antarctica. Non-systematic meteorite finds (meteorites that have been discovered by accident by non-professional meteorite searchers) are far more abundant in the United States than any other

RIGHT Part of the Barwell meteorite, which fell on Christmas Eve 1965, smashed through a window as it landed.

country. This is largely because of the work of a 20th-century meteoriticist, Harvey Ninninger, who travelled the country teaching farmers and land workers how to identify meteorites.

History of meteoritics

Europe provided the initial stage for the modern acceptance and understanding of meteorites. There are many folklore reports of meteorite falls in pre-Enlightenment times. Perhaps the most famous example is the Ensisheim meteorite, which fell in France in 1492. Under orders from King Maximilian of Germany, who ruled the area at that time, it was placed first in the local church, and then moved to the town hall, where it remains today. Despite the existence of this meteorite, the intellectuals of Europe remained sceptical of the idea of rocks falling from the sky until the end of the 18th century. Although the existence of meteors was universally accepted, reports of resulting meteorites were typically dismissed.

Scientists were unwilling to believe in a phenomenon for which no witnesses of suitably high social rank could be found, and for which no reasonable scientific theories could be offered.

A turning point was reached in 1794 on the publication of a short book by the German physicist Ernst Chladni. His research into meteors had led him to the conclusion that they were caused by extraterrestrial objects entering the atmosphere, which could sometimes survive their fall to Earth. His book was treated with scorn and incredulity by many of his colleagues, despite many witnesses reporting a fall of a large shower of stones in the Tuscan city of Siena in the same year as the book's publication. Chladni did not have too long to wait to be vindicated. In 1795 a chondrite fell on farmland in northern England, near Wold Cottage, Yorkshire. The owner of Wold Cottage, a Mr Topham, took the rock to a chemist, Edward Howard who discovered that its chemical composition was in many ways similar to that of

RIGHT Harvey Ninninger using a metal detector in the 1930s to search a field for meteorites in West Texas, USA, probably in the Plainview strewnfield.

other rocks that were reported to have fallen from the sky. What was more, these rocks all contained nickel dissolved in iron – a characteristic that was not observed in any terrestrial rocks. He concluded that a common extraterrestrial origin of these samples was plausible.

Any remaining doubts among the scientific establishment that meteorites were truly a real phenomenon were quashed when an impressive meteorite shower of thousands of stones was observed in L'Aigle, France in 1803. Faced with such overwhelming evidence, the intellegentia finally accepted that rocks from space could occasionally fall to Earth, and the science of meteoritics was born.

Hot desert collection

To date, many thousands of meteorites have been found – mainly during professional searches – in the desert regions of the world. Some parts of the Sahara Desert have proved particularly productive for meteorite searches, especially the Dar al Gani, Hammadah al Hadara and Adrar areas. Meteorites have also been recovered in other desert regions around the world, including the Australian Nullarbor, the Namib of southwest Africa, and the Chilean Atacama. The best places for desert collection are flat areas (without any contributions of exotic rocks and pebbles from local mountains and hills) that have remained arid for thousands of years. A pale-coloured local rock is helpful since it provides a good contrast to the typically dark-coloured extraterrestrial material. Systematic collection searches can be performed either on foot or by vehicle, the former providing a more comprehensive search of a small region, and the latter yielding only larger meteorites.

Antarctica

In 1969, a group of Japanese geologists working in Antarctica came across an unusual find. They discovered several meteorites in a single area, and

BELOW LEFT A small meteorite found during a systematic search of part of the Nullarbor region, Western Australia.

BELOW The Meteorite Hills region of Antarctica, a site in which around a thousand meteorites have been found.

examination showed that these meteorites were different from each other, rather than being fragments of a single fall. Since then official expeditions, led by Japanese, American and European groups, have successfully found tens of thousands of meteorites. There are now more meteorite finds from Antarctica than all the other continents put together.

Antarctica has proved the most fruitful continent of all for meteorite collection for three main reasons. Firstly, the plateau of Antarctica is a desert, the driest place on Earth. It is also the coldest continent. Because of the climate, any meteorites that land in Antarctica are likely to remain preserved, deep frozen, for up to several millions of years. Secondly, on the Antarctic ice, meteorites are easy to spot: in some places where there are no other sources of rocks and pebbles, the only dark objects on the ice are likely to be meteorites. Without the distraction of other rocks, it is possible to find unusual types of meteorite that may go unrecognised in more rocky regions. The final reason that Antarctica has yielded many meteorites for scientists is that the movement of the glaciers can act to concentrate meteorites in certain places. In some regions of the continent, the path of a glacier towards the ocean is blocked by mountains, forcing parts of the glacier into a 'cul de sac'. The ice in this region is stripped away (ablated) by the wind, leaving a residue of the rocks contained within the glacier. This process concentrates any rocks, including meteorites, that were being carried within the ice. The ablating ice regions are called 'blue ice' as they have a pale bluish tinge. The blue colour can be observed from satellite photographs, and provides a good way of finding new potential sites for meteorite searches.

Under the Antarctic treaty, Antarctica cannot be used for mining for financial gain. All geological samples taken from Antarctica can be used only for scientific research. Therefore, all Antarctic meteorites recovered after the treaty was signed are kept within scientific institutions and cannot be bought or sold.

BELOW Diagram to show how meteorites may be concentrated in certain parts of Antarctica. As glaciers move towards the ocean they carry the meteorites with them. But if their path is blocked, the ice is stripped away by the wind exposing the residue of meteorites.

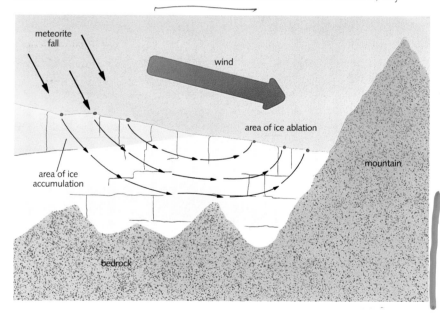

meteorite fall

wind

area of ice ablation

area of ice accumulation

mountain

bedrock

Asteroids and comets

The predominant sources of meteorites are the asteroids. One small group of meteorites comes from the Moon (see p. 37), and another from Mars (see p. 39). It is also possible that meteorites could come from comets, or the Kuiper Belt.

What are asteroids?

Asteroids are small bodies of solid material remaining from the earliest epoch of Solar System history. Over 25,000 have been identified and named so far, although there are probably several hundred thousand more awaiting discovery. Most of these minor planets, or planetesimals, orbit the Sun in a belt between Mars and Jupiter. They form the boundary between the inner rocky planets of the Solar System and the outer gas- and ice-rich planets. Asteroids exhibit a variety of orbits, some of which approach or cross that of the Earth. Occasionally these near-Earth objects (NEOs) actually hit the Earth and past collisions have resulted in significant changes both to the Earth's surface (with the formation of craters) and its biodiversity (with impacts leading to gross evolutionary change). Images from three asteroid fly-by encounters show that asteroidal surfaces are heavily cratered, implying a long collisional history for these bodies.

Asteroids never seem to have been part of a single planetary parent: they were prevented from aggregating together by the gravitational influence of Jupiter, which acted to keep the bodies apart. Jupiter is still the major factor that dominates the structure and dynamics of the Asteroid Belt. The asteroids are not distributed in a regular and even

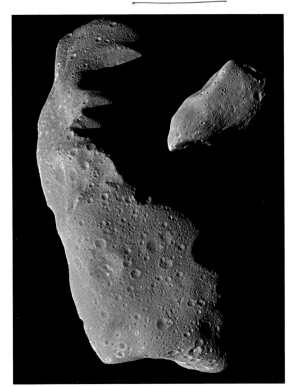

LEFT Photomontage of two asteroids: Ida (left) is about 30 km long by 10 km wide. Gaspra (right) is about 17 km long. Their irregular shapes and cratered surfaces show their long history of collisions.

RIGHT Distribution of asteroids in the inner Solar System.

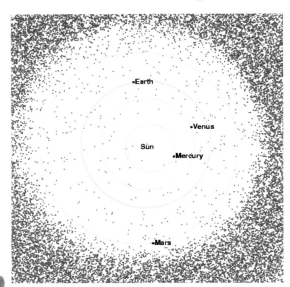

Structure of the Solar System

Our local star, the Sun, dominates our Solar System, making up more than 99% of its total mass. Orbiting around this star are nine planets. The inner four – Mercury, Venus, Earth, and Mars – are called terrestrial planets, as they have a hard, rocky surface. Beyond Mars lies the Asteroid Belt, a zone particularly rich in small rocky bodies, and probably the original source of most of the meteorites that land on Earth. Past the Asteroid Belt lie the gas giant planets – Jupiter, Saturn, Uranus and Neptune. And around and beyond Neptune is another belt of small bodies, this time more icy than rocky in composition. This belt is called the Kuiper Belt, after the astronomers that predicted its existence in the 1950s. The largest member of this belt is big enough that it has been recognised as our ninth planet – Pluto.

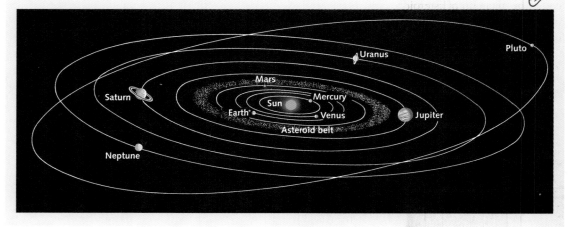

pattern across the belt – there are clusters and gaps. If an asteroid strays into one of these gaps, it becomes subject to gravitational disturbance from Jupiter, and is ejected from the belt. Many asteroids ejected from the belt are thrown out of the Solar System, or fall inward to the Sun, but a few crash into planets. Fragments of asteroids that hit the Earth are called meteorites. Meteorites have different compositions from each other, which probably reflect differences in the orbit of their parent asteroid, suggesting the Asteroid Belt is zoned in composition.

Classification of asteroids

Much has been learnt about asteroids by remote observations, mainly from measurement of their colour, or spectroscopic signature (the way in which they reflect sunlight back from their surface). Spacecraft encountering asteroids on their journey to other planets have also gathered data. The probe Galileo flew past Asteroid (951) Gaspra in 1991, as it travelled to Jupiter and the NEAR (Near Earth Asteroid Rendezvous) mission orbited Asteroid (433) Eros between February 2000 and February 2001, sending back high-quality, high-resolution images and spectra of the asteroids.

Asteroids are not all the same. They can be sub-divided into a variety of types on the basis of their spectral signature. Different signatures imply different compositions: rocky, metallic, rock plus metal or carbonaceous. There are at least 14 different asteroid groups, each with its own spectral characteristics. Asteroids are also divided into families, based on their orbits. Families tend to be made up of asteroids with the same composition, and are assumed to have formed from the break up of a larger body.

Relationship to meteorites

Spectroscopic data of asteroids have been compared with data obtained from meteorites. There are good matches between some asteroid types and specific meteorite classes, for example, the match between asteroids of spectral class C (the most abundant asteroid class) and some carbonaceous chondrites, a type of stony meteorite. However, the distribution of asteroidal types (as determined by remote spectroscopic techniques) is different from the distribution of meteorite types. This implies that the meteorites collected on Earth are not typical of the total asteroid population. Meteorites (originating from around 100 parent bodies) are a biased and incomplete sample of the asteroid population, representing only the fraction of the Asteroid Belt that has been involved in collisions and has ejected fragments into the inner Solar System.

There are some inconsistencies when trying to match meteorite types with potential asteroidal parents. For example, the largest group of stony meteorites (accounting for about 80% of all meteorites seen to fall), the ordinary chondrites, have no direct spectral match with a particular asteroid type. The second most abundant asteroid class is that of the S-type asteroids, which are divided into seven sub-groups. Asteroids in S(IV), exhibit spectra that are similar, but not identical to, those of H-group ordinary chondrites.

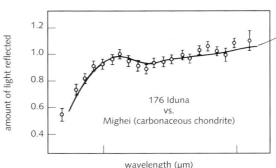

176 Iduna
vs.
Mighei (carbonaceous chondrite)

amount of light reflected

wavelength (µm)

LEFT Comparison between the spectral signature ('colour') of an asteroid (circle with bars) with a stony meteorite (solid curve).

FAR LEFT Asteroid (433) Eros was the target of the NEAR mission. It is a stony asteroid about 33 km long and 13 km wide.

Naming of asteroids

The largest asteroid, and the first to be discovered (in 1801), was Asteroid (1) Ceres. The next nine asteroids to be observed and named were Pallas, Juno, Vesta, Astraea, Hebe, Iris, Flora, Metis and Hygiae – all goddesses from Greek and Roman mythology. The tradition of giving female names lasted up until around the end of the 19th century, by which time over 500 asteroids had been named. In 1892, by the time Asteroid (332) Siri had been named, asteroids were being discovered so frequently that it became necessary to introduce a more systematic naming scheme. Asteroids were given a number-letter combination – for example, 1893 E was the fifth asteroid discovered in 1893 – as well as a name. It soon became apparent that this system was inadequate, with more asteroids being discovered than letters in the alphabet.

From 1925, a consistent numbering style has been used for newly discovered asteroids. They are assigned a number-letter combination that starts with the year of discovery, followed by letters to indicate the half-month period and the order in which the asteroid was discovered that month. Names are still given to asteroids, but have to be approved by the International Astronomical Union (IAU), which has strict guidelines for naming asteroids. Most up to Asteroid (10000) Myriostos (the Greek for 'ten thousand') have names, but asteroids are now being discovered at such a rate that it is hard for the IAU to keep up with naming them. A search through the list of asteroid names shows the diverse interests of astronomers. Asteroids have been named in addition to their official name, after musicians, such as (1034) Mozartia and (1815) Beethoven, as well as (4147) Lennon, (4148) McCartney, (4149) Harrison and (4150) Starr; artists, including (6676) Monet and (4511) Rembrandt; writers, such as (2985) Shakespeare and (3047) Goethe; scientists, (697) Galilea, (662) Newtonia, (8050) Isaacnewton, (1991) Darwin, and (2001) Einstein; plus other miscellaneous people, such as (9007) Jamesbond and (4731) Monicagrady.

The differences in spectra probably result from weathering in space, the process by which the surface of an asteroid suffers physical and chemical changes as it is bombarded by solar wind and asteroidal debris. Space weathering can alter the mineralogical properties of asteroidal surfaces, obscuring links between meteorites and their parent bodies.

Comets

The word comet derives from the Greek word 'kometes', meaning 'long-haired'. Early astronomers chose this description because the most notable

BELOW Halley's Comet as it appeared in February 1986. The bluer of the two tails (uppermost) is made up of gases, whilst the lower tail, illuminated by reflected sunlight, is the dust tail.

feature of a comet is its long tail that streams away from the bright head, or nucleus. The cometary nucleus is a small, fairly compact object, only a few kilometres in length. It is surrounded and hidden by a cloud of gas and dust called the coma. When the European Space Agency's Giotto mission photographed Halley's comet in 1985, it was the first time that the irregular shape of a cometary nucleus had been observed. Cometary nuclei are ice-rich bodies containing dust (the 'dirty snowball' model, proposed by astronomer Fred Whipple in 1950-51); they are assumed to be highly porous with voids and cracks from which gas can escape. The nucleus is dark, indicating that it probably has a dusty crust. The dust is not just composed of silicate minerals, but has a high percentage (up to 20%) of organic material mixed in with it. Results from observation of Halley's comet and comet Hale-Bopp have refined Whipple's 'dirty snowball', to a model in which a cometary nucleus appears to be a highly porous assemblage of silicate grains coated by organic compounds and bound together by ices.

A comet actually has two tails, one of dust and the other of plasma (charged particles). The tails point away from the Sun, as the particles are pushed away by solar radiation pressure. The dust tail can extend for millions of kilometres away from the nucleus; the accompanying plasma tail is usually more than ten times longer. The curved dust tail is made of particles shed from the cometary nucleus. Sunlight reflected off the dust gives the tail a yellowish colour. The grains are less than 1 micron across, and are predominantly made from carbon-rich compounds layered on top of silicates. In contrast, the bluer plasma tail is made up of ions and electrons. It is a long, straight tail, the structure of which is controlled by magnetic fields within the tail and the interaction of these fields with the solar wind magnetic field.

As a comet approaches the Sun, the icy nucleus is heated, ices vaporise and the coma forms. Dust grains trapped in the ice are released and stream away from the nucleus. A comet can lose up to 1% of its mass each time it passes close to the Sun – in the case of Halley's comet, that is around 100 tonnes

BELOW Diagram of a comet, showing its nucleus, the coma and tail.

BOTTOM Comet Hale-Bopp which appeared in the night skies of the northern hemisphere in 1997.

plasma (ion) tail

nucleus

coma

dust tail

per second. The coma first appears when the comet is about 3–5 AU from the Sun, when ices start to vaporise (one AU, or Astronomical Unit, is the mean distance from the Earth to the Sun, about 150 million km).

Where do comets come from?

There are two types of comet: long period and short period. Long-period comets approach the Sun from random directions and travel along elliptical orbits around the Sun that take more than 200 years to complete (for example, comet Hale-Bopp, which last appeared in 1997, has an orbital period of about 3000 years). Short-period comets approach the Sun at fairly shallow angles and have much shorter orbits (for example, Halley's comet orbits once every 76 years). The main reservoir of long-period comets is the Oort Cloud at the outermost edge of the Solar System.

LEFT An impression of the Leonid meteor shower of November 1799, so-called because the meteors appear to radiate from the constellation Leo.

Kuiper Belt objects

The Kuiper Belt is a reservoir of small bodies that lies beyond the orbit of Neptune (the outermost gas giant planet), at a distance of between 30–50 AU from the Sun. Its existence was proposed independently by Kenneth Edgeworth in 1949 and two years later by Gerald Kuiper in 1951, although the first Kuiper Belt object (KBO) was not observed until 1992. It has only been possible to study KBOs in the last decade, since advances in instrumentation have enabled the detection of these faint and dim bodies. It is estimated that there are about 35,000 KBOs. They have characteristics in between those of asteroids and comets: almost circular orbits, dark surfaces, presumably ice-rich, up to about 300 km in diameter. Pluto, which has never fit comfortably into its

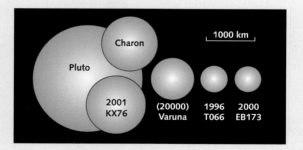

classification as a planet, might also be a KBO, but much larger than any of the others with a diameter about 2300 km. Because they have never been close to the Sun, KBOs have never been heated, so they are probably even more pristine in their composition than comets.

ABOVE LEFT Sizes of the largest known KBOs relative to Pluto and its moon, Charon.

Cometary missions

Recognition that comets are probably the most primitive bodies in the Solar System has resulted in a series of space missions to comets. The most successful to date was the European Space Agency's Giotto mission to Halley's comet in 1985. The Giotto spacecraft took high-resolution images of the cometary

RIGHT The nucleus of Halley's comet taken by the Giotto spacecraft. The nucleus is about 16 km by 8 km.

nucleus, discovering its very irregular and elongate shape. NASA's STARDUST mission (launched in February 1999) is currently in progress. This is an ambitious mission that aims to pass only 100 km away from the nucleus of comet 81P/Wild2 in 2004. As well as photographing the nucleus, it will collect samples of dust from the coma. The dust will be returned to Earth in 2006. NASA will launch two further cometary missions in 2002: DEEP IMPACT will approach comet p/Tempel1, and fire a copper projectile into its nucleus, excavating material and exposing the comet's surface. CONTOUR is NASA's cometary nucleus tour. It will make close fly-bys of at least two (and possibly four) cometary nuclei between 2003 and 2008. ESA's Rosetta mission (launch 2008) will rendezvous with comet 46P/Wirtanen in 2011, releasing a lander on to the nucleus. Instruments on board the lander will measure the composition of cometary dust, ice and gas.

Although this body has never been observed directly, its existence is inferred from observations of cometary orbits. Short-period comets were once thought to be long-period comets that had been perturbed by the giant planets into different orbits. However, it is now recognised that short-period comets emanate from the Kuiper Belt (see box).

Meteor showers

Each time a comet approaches the Sun it loses dust. Eventually, gravity and radiation pressure smear the dust out, until it occupies the whole of the comet's orbit. If the Earth's orbit crosses the cometary orbit, then the cometary debris produces a meteor shower in the Earth's atmosphere. There are several such showers, for example, the Leonids (associated with comet 55P/Tempel-Tuttle; maximum intensity around November 17) and the Perseids (associated with comet 109P/Swift-Tuttle; maximum intensity around August 12). The dates of the showers are predictable, although shower intensity for a particular stream varies, depending on where the comet is in its journey, and whether the stream has been 'recharged' with debris from a recent passage. For instance, in the Leonid shower the number of meteors was expected to be highest in November 1999, following the 33-year period of comet 55P/Tempel-Tuttle. It is important to note that no particles are recovered from a meteor shower as all the material is burnt up in the Earth's atmosphere.

Relationship between asteroids and comets

Asteroids have been changed to varying extents since they formed, from heating and from water flowing through them. Comets, on the other hand, are believed to be primitive, with the dust fraction more closely representing that from which the Solar System originally formed. The distinction between comets and asteroids is less clear cut than has been believed: although comets have active comas (within the inner Solar System) and less circular orbits than asteroids, cometary nuclei can become dormant and their orbits can evolve into similar paths to those of asteroids. Likewise, some asteroids – especially those in the outermost part of the Asteroid Belt – presumably still contain ice locked below their surface. It is possible that if an asteroid approaches sufficiently close to the Sun, it might develop a coma. This happened to Asteroid (4015) in 1979. A check of its orbit showed that it was the 'lost' comet 107P/Wilson-Harrington, discovered in 1949. Kuiper Belt objects (KBOs) are intermediate in their properties between asteroids and long period comets.

Unmelted meteorites and planet formation

As discussed on p. 5, meteorites are divided into three main types: stones, irons, and stony-irons. Stony meteorites are further subdivided into the unmelted chondrites and the melted achondrites.

Chondrites are thought to have formed at the same time and from the same material as the inner rocky planets of our Solar System. Studying them can tell us about how the planets were made and about the materials that originally formed our own planet.

Chondrites are a jumbled mixture of materials; most of the components are usually a few millimetres across or smaller. The major constituents are chondrules – rounded, millimetre-sized, once-molten objects that formed during a fast heating event. (Chondrules are named from the Greek word 'chondros' meaning 'seed' or 'little grain'.) Chondrites also contain shiny flecks of iron-nickel metal, plus pale-coloured patches of calcium and aluminium rich inclusions (CAIs). The fine-grained matrix surrounding these objects is made up of fragments of chondrules, along with organic material and ancient presolar grains (see p. 28).

Chondrites are divided into three main classes: ordinary, carbonaceous, and the rare enstatite chondrites. These three types may have formed at different distances from the Sun. Ordinary chondrites, as the name implies, are the most common type of meteorite. They account for around 90% of all meteorite falls.

Carbonaceous chondrites are much rarer, but perhaps can tell us even more about the origins of the Solar System. They have a composition that is richer in volatile substances (such as carbon compounds) than the ordinary chondrites. In fact their composition is extremely similar to that of the Sun, minus the hydrogen and helium that make up the vast majority of solar material. For this reason they are thought to represent the composition of the inner planets within the Solar System, and their composition is used by geochemists to compare to terrestrial rocks.

Enstatite chondrites are notable for having very oxygen-poor compositions, so much so that elements that usually bond with oxygen in other

BELOW LEFT The Parnallee ordinary chondrite containing many chondrules.

BELOW RIGHT Except for gases such as hydrogen and helium, the chemical composition of chondrites and the Sun is very similar.

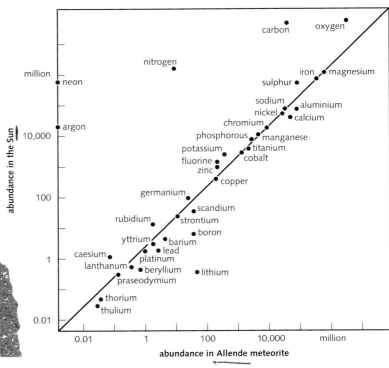

meteorites (and on Earth) have formed compounds with sulphur. In addition to the three main types of chondrite – ordinary, carbonaceous and enstatite – there are also several chondrites that do not fit very well into any of the known groups.

As well as differing in chemistry, the different chondrite groups can be distinguished from each other by their oxygen isotopes. Oxygen has three isotopes: oxygen-16 (the most abundant), oxygen-17 and oxygen-18. The proportion of these three isotopes varies between meteorites. Meteorites that differ in oxygen isotopic composition either formed on different parent bodies or formed in different parts of the same parent body.

Planet formation

The mechanism by which planets formed can be investigated in three ways: by looking at rocks that

formed very early in our Solar System's history, by observing young stars to look for signs of planet formation, and by theoretical models of planet formation. All three approaches have contributed to what we know of our planet's origins. To understand what meteorites can tell us about the formation of the planets, we must first look to theoretical and observational information about planet formation.

A history of planet formation theories

Since Copernicus' discovery that the planets orbit the Sun, there have been three major theories about how the planets formed. One of the most popular theories, first postulated by astronomer James Jeans (1877-1946) was the ejection hypothesis, which states that the material planets were formed from was ejected from the Sun as a

RIGHT Formation of the Solar System

1 A turbulent cloud of interstellar dust and gas collapses.

2 The dust and gas cloud forms a spinning disk.

3 The temperature and pressure of the disk increases towards the middle. Eventually, the temperature is hot enough for fusion of hydrogen into helium to begin. The central star is born.

4 The remaining dust and gas clump together, gradually sweeping up all the debris into planets.

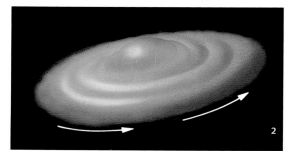

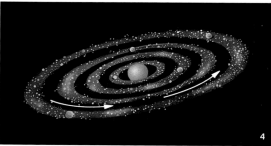

filament of gas following a collision with a giant comet or another star. The theory was later rejected because planets are relatively enriched in deuterium and lithium, which would be destroyed in a star. A second major theory, developed in the 1930s, was the capture hypothesis, which states that passing material was captured by the early Sun's gravitational pull. Although the capture hypothesis has never been disproved, recent astronomical observations make the third, and oldest theory – the nebula hypothesis – more attractive. By the nebula theory, the Solar System formed from a rotating, flattened disc. This idea was first proposed by the Prussian astronomer Immanuel Kant in 1755, and further developed by Laplace in 1796. It remains the most widely accepted theory of planet formation to this day, because it is compatible with both the observational evidence and evidence from meteorites.

The modern-day nebula theory pictures the Sun and planets forming at the same time, from the same dense cloud of interstellar dust and gas.

As most of the cloud, collapsed by gravity, swirled inwards to form the Sun, the remains of the infalling cloud began to form a rotating disk, called an accretion disk or solar nebula, around the central star, from which the planets were born.

Observational evidence: protoplanetary and debris discs

Most very young stars – less than about 10 million years old – have discs of dust and gas, called protoplanetary discs. These discs become less and less common around progressively older stars, as they dissipate or form planets. The dust is a mix of silicates and carbonaceous material – much like our own Solar System at the time of its formation – and the gas is mostly hydrogen and helium.

Around 15% of nearby older stars also have discs around them. These discs have proportionally high amounts of dust, and low amounts of gas. One of the best known examples is *Beta pictoris*. Since dust is expected to fall into the star over a short timescale, these dusty discs

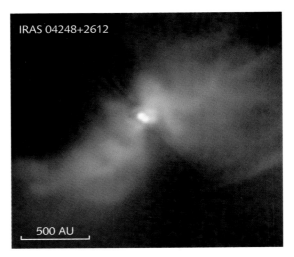

CoKu Tau1

500 AU

IRAS 04248+2612

500 AU

LEFT Young stars imaged using new-generation telescopes provide clues about the origins of our own Sun and Solar System, and show how the planets may have formed from a dusty disk surrounding the young Sun.

around old stars are probably continually replenished with new dust from asteroids and comets. Some of the discs are ring-shaped, with a hole close to the star. What is in the hole? Paradoxically, telescopes enable us to detect the tiny dust grains even though at the same range planets, asteroids and comets remain invisible to us. The inner holes could be full of planets. In fact, our own Solar System, if viewed from a neighbouring star, would also look like a star surrounded by empty space in place of the planets, but clearly showing the dust ring of the Kuiper Belt.

Formation of terrestrial planets: the story from meteorites

Much of our knowledge about the formation of terrestrial planets – the four inner, rocky planets – comes from meteorites. Meteorites reveal a complex picture of planet growth. The first stage is the coagulation of dust grains, that are initially only a millionth of a metre across or less. Dust coagulation was aided by heating events that allowed the dust to melt and stick together to make chondrules. These objects then stuck together, perhaps using organic material as glue, to form metre to kilometre-sized boulders, called planetesimals. Once the planetesimals reached a certain size – greater than a kilometre or so – they began to have a significant gravitational field. This made their growth much easier and quicker, as other planetesimals were attracted to them and their growth progressed by impacts of these relatively large bodies. Although this broad outline of how planets formed seems to fit most of the features seen in meteorites, deciphering the history recorded in meteorites to understand the

details of planet formation proves much more difficult. Some important questions remain. How did chondrules form? What was the heat source for early planets? How long did the process of planet formation take? Ongoing studies of meteorites are helping to address these questions.

Components of chondrites
Chondrules

These are near-spherical, typically millimetre-sized stony (silicate) objects, that formed by a sudden, flash heating event at temperatures of over 1400°C followed by fast cooling and solidification of the resulting molten droplets. Chondrules are mostly made up of the magnesium, iron and silicon-rich minerals olivine and pyroxene. The subject of exactly how chondrules formed, and what was the mysterious heat source that produced these objects early in Solar System history, is one of the

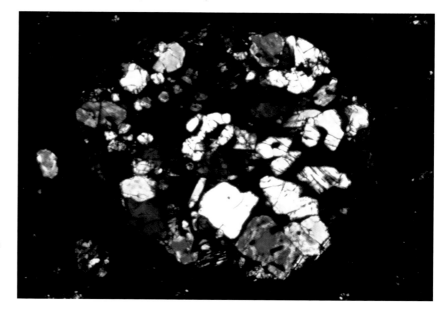

BELOW A chondrule from the Cold Bokkeveld carbonaceous chondrite, viewed in an optical microscope. The chondrule is about 0.5 mm across.

long-standing mysteries in meteoritics. It is important to find the answer to this question, since this tells us how the very first stage of planet formation, the accretion of tiny dust particles into larger millimetre to centimetre-sized balls, took place. By looking at the chemistry and texture of chondrules we can place a few constraints on how they formed. For example, many chondrules appear to contain 'chondrules within chondrules' suggesting that they experienced a high-temperature heating event more than once. The main theories are ① formation in a shock wave in the early Solar System, ② formation in a jet from the Sun, ③ formation by collision between planetesimals.

The shock-wave theory argues that in the early Solar System there would have been waves of high pressure passing through the inner region where the planets and asteroids formed. The shock of being hit by this wave of pressure would be enough to heat and melt dustballs and cause them to form chondrules. The second theory argues that the chondrules formed extremely close to the Sun. Particles spiralled inwards towards the early Sun, and most material ended up flowing into the growing Sun. However, changes in the magnetic field around the Sun cause material to rise up in big jets. As the grains were lifted, they were exposed to the full intensity of the sunlight which heated them up and caused chondrules to form. The jets then deposited the chondrules onto the accretion disk at the distance that the asteroids and planets were forming. Evidence that this process occurred is found in observational astronomy – young stars are observed to have jets streaming out. The main evidence for the third theory, planetesimal collisions, is that some chondrites contain rare rock fragments that have undergone melting themselves. It is argued that these rock fragments came from planets that must have existed before the chondrites formed. Chondrules may then have formed as a product of high-energy collisions between planetesimals just tens or hundreds of kilometres in diameter.

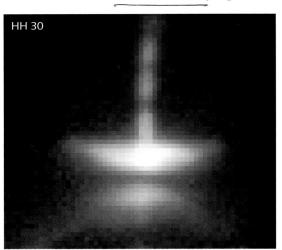

HH 30

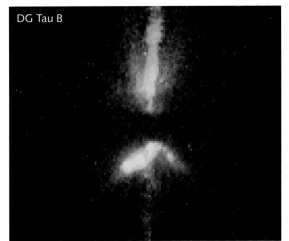

DG Tau B

RIGHT These images are young stars obscured by a dark disk of dust. Because of changes in the magnetic field of the star violent jets of material are created which can be clearly seen here.

Metal

One thing that best distinguishes chondrites from terrestrial rock, is that they almost always contain flecks of metal – a mixture of iron and nickel. This makes chondrites slightly attracted to a magnet, and provides a good test for meteorites. The presence of metal is extremely important. It implies that the rock has not been altered since the earliest times in the Solar System since the metal has not melted and sunk to the core of its parent asteroid – as has happened on Earth. The quantity of metal in chondrites is very variable – from 2% to 20% – and metal content provides an important way of classifying chondrites of different types.

CAIs

In February 1969 a shower of stony meteorites fell near the village of Pueblito de Allende, Mexico. The meteorite turned out to be of the rare carbonaceous chondrite variety, and many of the stones were quickly retrieved and eagerly studied by laboratory groups all over the world. One immediately obvious and unusual feature of the Allende meteorite is that on each surface, millimetre to centimetre-sized, irregularly shaped, pale coloured inclusions are clearly visible. These are calcium aluminium-rich inclusions (CAIs). CAIs have a composition that is subtly different to chondrules; they are typically much more

RIGHT The Allende carbonaceous chondrite, partly covered in jet-black fusion crust. This contains numerous white inclusions called CAIs. This stone is about 10 cm across.

enriched in the high-temperature elements calcium and aluminium, and contain less abundant silicon and magnesium and more volatile substances. The ages of several CAIs have been calculated and found to be the oldest known solids in the Solar System – dating from around 4560 million years ago. They are quite common in carbonaceous chondrites, but extremely rare in ordinary and enstatite chondrites. They are also different to chondrules in their oxygen isotope composition. The formation mechanism of CAIs is unknown, but it may be similar to that of chondrules, although CAIs seem to have been heated for longer, and cooled more slowly. An important difference between the two types of objects is that not all CAIs formed from molten droplets. A few CAIs have fragile, intricate textures like snowflakes, suggesting that they condensed directly from a vapour to a solid without ever being molten.

BELOW LEFT A CAI from the Leoville carbonaceous chondrite. It is 2 cm across.

BELOW RIGHT A cluster of millions of diamonds, separated from the Allende meteorite. Some scientists believe these are presolar.

Presolar grains

The matrix of meteorites contains some tiny crystals of very resilient minerals, such as diamond, silicon carbide, corundum, silicon nitride and graphite. The isotopic composition of these grains is completely different to the composition of everything else in their host meteorites, or indeed anything on Earth or in the rest of the Solar System. These 'presolar' grains are so named because they are thought to have existed before the Solar System formed. They are ancient minerals that originally formed around stars, such as red giants or supernovae. Stars such as these were our ancestors, producing the elements from which the Solar System formed. The grains produced around these stars floated in interstellar space before being swept up and incorporated into our newly forming Solar System. Presolar grains can provide unparalleled opportunities to perform astrophysics in the laboratory, to learn about the element- and

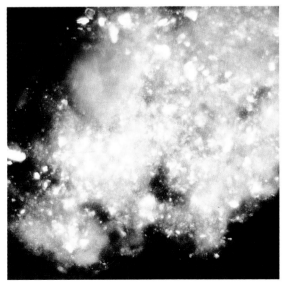

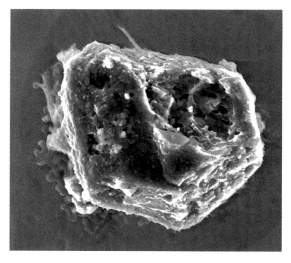

RIGHT A presolar grain of silicon carbide, about 1μm across, from the Murchison meteorite.

grain-forming processes occurring in and around other stars. Studies of presolar grains have told us that the Solar System is made up of some material from a supernova as well as material from several red giant stars. All in all, dozens of stars separated by billions of kilometers all contributed to our Solar System, ultimately to end up in meteorites that sit side-by-side in a single room.

Organic compounds

Some meteorites contain organic compounds, made up of the elements carbon, hydrogen, oxygen, and nitrogen. These compounds, found most abundantly in carbonaceous chondrites, are the main constituents of living things. They are thought to have formed abiogenically (i.e. no life was involved in their formation) in interstellar space prior to the formation of the Solar System, within the solar nebula, or in the asteroid parent. Most likely their formation involved a mixture of all these environments. Meteorites containing organic material probably fell to Earth before life flourished here, and may indeed have been an important first step to the origin of life.

Heating of young planetesimals

Nearly all chondrites, and certainly all melted meteorites (see p. 26), show evidence of having been heated since they formed into a planetesimal. In the case of achondrites and iron meteorites, this heating event was clearly very severe, severe enough to cause the rock to melt. In the case of chondrites, the extent of their alteration since formation is very variable and provides a secondary classification scheme for these meteorites. Chondrites are classified by number from 1 (unheated but water damaged) to 7 (very severely heated). A few rare meteorites seem to have experienced very little heating since they formed and apparently remain almost unchanged. Others experienced minor heating, but have nevertheless been extensively altered by water on their parent bodies. However, the majority of meteorites have textures suggesting that they have suffered changes brought about by heat. Heating has altered the mineralogy of the rocks and has caused the constituent elements to move around and homogenise. Some of the heat in planetesimals is simply gravitational: as the components of the chondrites collapsed to form a larger body they released heat. However, this process cannot account for much of the heat required to produce the effects that are observed. We must look to a different mechanism, and the most likely heat source appears to be from the decay of radioactive isotopes with extremely short half-lives (that is, unstable isotopes that decay quickly, producing energy).

Isotopic studies of meteorites show that they did contain short-lived isotopes when they formed. The most important heat sources may have been the isotopes aluminium-26, calcium-41, and iron-60. We know that these isotopes were once active inside meteoritic material because today the meteorites contain excessive amounts of the isotopes' decay products (for example, magnesium-26, a decay product of aluminium-26). The short-lived isotopes may have formed either in a stellar event – such as a supernova that must have happened very soon before the formation of the Solar System so that the isotopes were still active – or else they may have formed close to the young Sun and were transported to the asteroid – and planet-forming region in jets. Once the origin and initial distribution of short-lived isotopes is understood, they can be used to indicate timescales in the early Solar System. Meteorites containing a large amount of decay products must have contained a large amount of radioactive nuclides when they formed. Meteorites that contained large amounts of radioactive nuclides are likely to be older than ones that contained small amounts (when the isotopes may have already decayed away). For example, the work of scientists Alex Halliday and D-C Lee on the short-lived isotope hafnium-187 has been used to infer that the planet Mars formed relatively quickly – within a few tens of millions of years of the start of the Solar System – because isotopic studies show that Martian rocks contained this isotope when the planet formed. In contrast, the Earth formed much later when all the hafnium-187 had already decayed away.

How old are meteorites?

There are three ages important in the history of each meteorite:

1. The age when the constituent minerals originally formed (formation age);
2. The time the meteorite spent separated from its parent body in transit to the Earth (cosmic ray exposure age);
3. The time it has spent since falling to Earth (terrestrial age).

Each of these three stages can be measured.

RIGHT An optical microscope image of the Parnallee (Type 3) chondrite which has experienced little heating. The chondrules are clear and well-defined. Field of view is 5 mm.

FAR RIGHT An optical microscope image of the Barwell (Type 6) chondrite. This meteorite has experienced a significant amount of heating so that the chondrules are not so well defined as for Parnallee. Field of view is 5 mm.

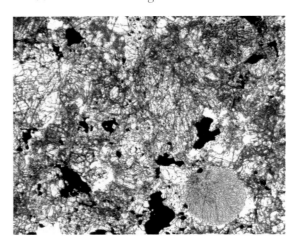

Formation age

This is the time since the formation of the meteorite's parent body. In the case of melted meteorites, this means the time since the minerals cooled from a molten state to their current solid state. In the case of unmelted chondrites, this can either mean the age of the individual constituents of the meteorite or else the age at which the constituents coagulated together. The formation age of all meteorites from the Asteroid Belt is around 4560 million years, the time of planet formation. Since then, the asteroids have been cold and inactive. Younger rocks are occasionally formed when asteroids collide together, producing enough heat to melt and reform their constituent rocks.

The ages are measured using radioactive 'clocks'. Elements sometimes have more than one isotope, nuclides of different masses to each other. While many isotopes are stable and can survive intact indefinitely, some are radioactive and break down to form different isotopes and elements. The rate at which these radioactive isotopes disintegrate is known, and so by measuring the amounts of both the parent isotope and its daughter decay products present in a rock, the age of the rock can be determined. The age determined is the time from when the rock was last hot enough for its constituent isotopes to be able to move around freely. The isotopes that are most commonly used for age dating of ancient Solar System rocks are rubidium-87 (which decays to strontium-87 with a half-life of 49 billion years), samarium-147 (which decays to neodynium-143) and various uranium isotopes that decay to lead isotopes. Most meteoritic fragments of asteroids yield similar ages, of around 4560 million years, older than any rocks that originate on Earth. Thus, the Solar System is believed to be 4560 million years old.

Cosmic ray exposure age

When a rock is in space, it is exposed to cosmic radiation. This radiation can interact with and damage the atoms in the rock, causing them to be altered into different isotopes e.g. iron-56 can be turned into radioactive manganese-53. Most of these new isotopes are stable and build up over time and are more abundant in rocks that have spent a long time in space. Measurements of cosmic ray exposure show that meteorites made of stone were typically on the journey to Earth for a few million or tens of millions of years. Meteorites made of iron are more robust, and the oldest one yet measured, the Deep Springs iron, has a cosmic ray exposure age of 2500 million years.

Terrestrial age

The time spent since the meteorite arrived on Earth is clearly well-known in the case of meteorite falls, where the arrival of the meteorite was witnessed. The terrestrial age can also be determined for meteorite finds. The way this is done is by using the measurement of cosmic ray exposure. Some of the isotopes formed by the cosmic rays are radioactive. By measuring how much they have decayed we can tell how long the meteorite has been on the surface of Earth, where it is shielded from cosmic rays by the blanket of the Earth's atmosphere. The most well-known isotope used for terrestrial dating is carbon-14. Terrestrial ages of non-desert meteorites are usually relatively short – up to about a thousand years. Hot desert finds can have ages of up to a

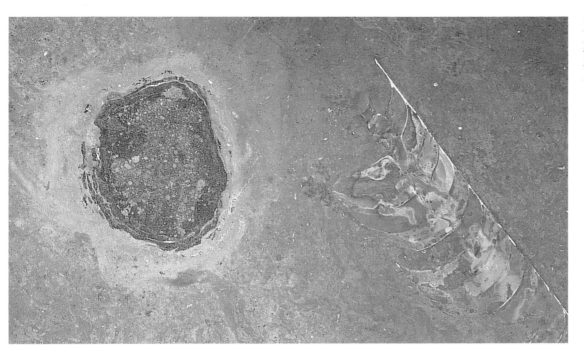

few tens of thousands of years, and meteorites from Antarctica are the oldest of all, having a terrestrial age of up to two million years.

There are around 40 examples known of fossil meteorites (embedded inside ancient rocks).

All of these are meteorites within Ordovician limestone in Sweden. The limestone surrounding the meteorites is around 500 million years old and so the meteorites fell to Earth approximately this long ago.

Melted meteorites and planetary evolution

On the surface of Earth many rocks were once molten, made of magma that cooled and solidified at or near the planet's surface to form 'igneous' rocks. We see similar kinds of rocks in meteorite collections – samples that have clearly once been extremely hot. These meteorites range in composition from those made of stone (which resemble terrestrial rocks), to rocks made of pure metal or mixtures of metal and stone. The metal-rich meteorites look quite different from any terrestrial material found close to the Earth's surface.

Achondrites are stony meteorites that have been melted. They are named to distinguish them from the chondrule-bearing chondrites, which have remained unmelted and preserve objects that were once floating in the solar nebula. For melting to have happened, they must have come from a parent asteroid of at least a few tens of kilometres in size – large enough to retain heat. Most achondrites, like the vast majority of all meteorites, come from asteroids. However, a few are thought to come from the Moon or Mars (see p. 37–45). Asteroidal achondrites show an enormous range of chemical composition and texture, reflecting the diverse nature of their sources.

Iron and stony-iron meteorites are also thought to come from large asteroids. Larger asteroids, like planets, differentiate to form a rocky outer layer and an iron-rich core. Iron meteorites are thought to be samples from the core of such asteroids, broken up during collisions. Some stony-irons, called pallasites, are composed of minerals surrounded by iron-nickel metal and are thought to originate from the core-mantle boundary of differentiated asteroids. Other stony-irons, called mesosiderites, formed during asteroidal collisions.

The structure of the Earth and other planets

As the Earth and other planets and large asteroids started to grow from chondritic material, they became hot due to gravitational energy collisions and radioactive heating. In cases where the planet was big enough, and contained enough radioactive isotopes, the planets started to melt. The nuggets of iron within them flowed and, being denser than the surrounding silicate rock, sank down to form a planetary core made mainly of iron. This process is called differentiation.

How do we know that these processes occurred naturally in planets? On Earth, we see only the

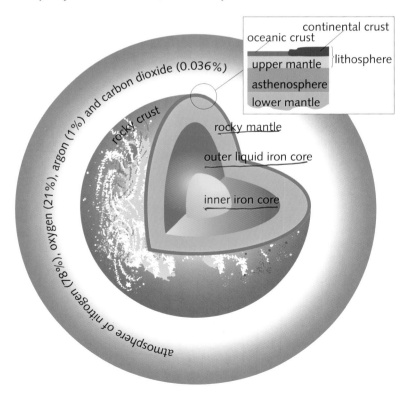

BELOW The structure of the Earth (not to scale).

continental crust
oceanic crust
upper mantle
lithosphere
asthenosphere
lower mantle

atmosphere of nitrogen (78%), oxygen (21%), argon (1%) and carbon dioxide (0.036%)

rocky crust

rocky mantle

outer liquid iron core

inner iron core

rocks from the planet's surface and near surface exposed for us to study. Even the deepest mines are only a few kilometres deep – a mere scratch compared to the diameter of the planet. We can learn about the internal structure of the Earth only by using geophysical techniques such as measuring how the Earth vibrates after an earthquake.

Meteorites that have experienced some melting can give us further insights into the structure and composition of the Earth. The composition of these meteorites span the range of the major layers that make up the Earth – from its iron core to its rocky outer layers.

Asteroidal meteorites are categorised according to their composition. The main division is simply between iron meteorites, composed of iron and nickel alloys, stony meteorites, and stony-iron meteorites (meteorites composed of a mixture of metal and stone).

Iron meteorites

Most iron meteorites were originally completely molten and formed from asteroidal cores (these are called magmatic irons). However, a few were not completely melted, and may have formed by impact processes rather than inside a core. Magmatic iron meteorites provide us with the best clues about the composition of the inaccessible Earth's core. Irons come from the centre of large asteroids that have been completely ripped apart by massive

ABOVE A piece of the Henbury iron meteorite, 28 cms across, which once formed part of an asteroid's core.

collisions. They are the only planetary core material that we can physically touch.

Irons are classified according to their chemical composition, particularly the abundance of nickel and also the abundance of elements present in trace amounts. Eighty-six per cent of iron meteorites can be classified into one of 13 groups. The others are unique ungrouped samples. Each group is thought to represent a different parent asteroid and so there are around 13 asteroids that are well-sampled. When the ungrouped meteorites are also considered, the number of iron meteorite parent bodies increases to over 60 – probably more than are sampled by stony meteorites. The high number of parent asteroids represented is probably related to the fact that more durable iron meteorites are likely to survive their passage through space compared with stony meteorites that can disintegrate during collisions. Iron meteorites are made up primarily of iron-nickel metal, with small inclusions of other minerals. Most iron meteorites contain 7–15% nickel. As the iron cooled from a hot liquid, it formed crystals of an iron-nickel alloy up to several metres across. On further cooling, to 600-300°C, two different alloys separated out of these large crystals, called kamacite and taenite. Kamacite typically contains about 95% iron and 5% nickel while taenite contains about 60% iron and 40% nickel. These two alloys separate into a distinctive criss-cross pattern called the Widmanstätten pattern, named after Count Alois von Widmanstätten, director of the Imperial Porcelain Works in Vienna. Count Widmanstätten discovered the pattern in 1808, at about the same time as a fellow scientist working in Naples, William Thompson. The Widmanstätten pattern reveals information about how the iron cooled from its initial molten state. Irons that cooled slowly, perhaps because they were in the very centre of a large asteroid and kept hot by a thick blanket of rock above them, had time enough to grow thick strands of kamacite and taenite and have a coarse Widmanstätten pattern. Irons that cooled more quickly have a finer pattern.

Interspersed among the metal grains in most iron meteorites are other minerals. The most common of these is troilite, an iron sulphide that formed dark, rounded nodules. Minerals rich in carbon and phosphorous are also quite common.

Stony-irons

There are two main types of stony-irons – pallasites and mesosiderites. The name pallasite comes from a German scientist, Peter Simon Pallas, who described the Krasnojarsk pallasite in 1772, before it was known that this sample was extraterrestrial. Pallasites are perhaps the most beautiful of all meteorites. They are made up of centimetre-sized green crystals, usually of a mineral olivine, surrounded by iron-nickel metal. These meteorites are thought to have originated at the core-mantle boundary of large asteroids. The metal they contain is very similar in composition to that from some iron meteorites, suggesting that they originate from the same type of asteroids.

Mesosiderites are an irregular jumble of metal grains and angular pieces of rock. The rocks are typically composed of pyroxene, plagioclase and olivine. They are thought to be the product of

LEFT Iron meteorites, when sliced open and etched with acid, typically show a distinctive criss-cross pattern called a Widmanstätten pattern. This slice is 15 cm across.

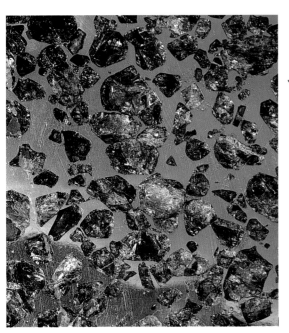

collisions in the Asteroid Belt between asteroids of different composition. Rocks that have a jumbled texture, like that seen in mesosiderites, are called breccias.

Melted stony meteorites

A few meteorites, called achondrites, are made mostly of stone but are not chondrites. They are igneous or once-molten in texture. Achondrites are divided into families or groups, and each family is unrelated to the other.

Howardites, eucrites and diogenites

The most common achondrites are from the 'HED association', standing for howardites, eucrites and diogenites. Diogenites are grey or green coloured meteorites made up mostly of the mineral pyroxene. These rocks formed as crystals settling to the bottom of a molten magma chamber.

LEFT The Esquel pallasite, composed of gem-quality olivine crystals embedded in metal.

BELOW LEFT The Estherville mesosiderite formed after a catastrophic collision between two asteroids.

BELOW A piece of the Stannern achondrite which is thought to have originated on the asteroid Vesta.

Eucrites are usually lighter in colour. They look very similar to a kind of terrestrial rock called a basalt and are mainly composed of the same minerals: pyroxene and plagioclase. Howardites are made up of fragments of eucrites and diogenites jumbled together. They probably formed at the surface of an asteroid, where constant bombardment from asteroidal collisions caused the surface rocks to become mixed.

The HED meteorites, although very different from each other in mineralogy and texture, have chemical similarities that strongly suggest they formed in the same asteroid. One indication is the similarity in their oxygen isotopes. We think we may even know which asteroid was the parent to these meteorites. The reflectance spectra of eucrites match the reflectance spectrum of one of the largest asteroids in the Solar System, 4 Vesta. So, it seems likely that the HEDs originated either from this one asteroid, or from the 'vestoids' – Vesta-like asteroids.

Aubrites

Unusually for meteorites, the aubrites are a light white-grey colour. They are mainly made up of enstatite, a magnesium-rich pyroxene and, unlike most other meteorites, do not contain much metal. These meteorites are thought to be closely related to the enstatite chondrite family of meteorites. In fact, rocks similar to aubrites can be made

artificially by heating and melting enstatite chondrites and removing most of their metal. The most likely history for aubrites is therefore that they started out very much like enstatite chondrites, but their parent asteroid was heated enough to melt and differentiate. Aubrites contain some minerals that are rare or unobserved in other meteorite samples. One example is found in the aubrite Bustee, which contains beautiful centimetre-sized chestnut brown crystals. These crystals are of a mineral called oldhamite, a calcium sulphide that is not found on Earth.

Ureilites

Ureilites are one of the oddest and least understood types of meteorites. They are made up of the minerals olivine and pyroxene, but they also contain veins of carbon-rich material in the form of graphite and diamond and they are extremely rich in noble gases. Diamonds on Earth mostly form at extreme depths, where the pressure of the overlying rock compresses carbon to its densest form. Diamonds found in ureilites apparently formed in a slightly different way. Because asteroids (from where ureilites originate) are so much smaller than Earth, even in their centre the pressure is not high enough to make diamonds. High pressures were involved, but in the case of ureilites this seems to be during a sudden shock event, perhaps when two asteroids collided at high speed. This shock event has partially obliterated the original form of the rocks and makes deciphering their origins even more difficult. How these rocks formed, combining as

ABOVE The Bustee aubrite, a light-coloured meteorite containing brown oldhamite crystals (seen here on the right-hand side of the image).

they do high temperature olivine and pyroxene with delicate carbonaceous material and volatile gases that must have always remained at low temperatures, is a mystery.

Acapulcoites and lodranites

Named after falls at Acapulco, Mexico, and Lodran, Pakistan, these are unusual for achondrites in that although they have a melted texture, their chemistry is very similar to that of chondrites. This implies that they apparently have not undergone differentiation. They appear to have formed by partial melting of chondrites at temperatures of around a 1000°C – not quite enough for complete melting to take place. In fact, in some cases residual chondrules can still be seen. These meteorites are often described as 'primitive achondrites' and they may provide a link between the chondrites and achondrites.

BELOW LEFT An optical microscope image of the FRO9023 ureilite which was found in Antarctica. The image is 1 cm across.

BELOW RIGHT Da Al Gani 400, a lunar meteorite found in the Sahara Desert.

Lunar meteorites

The idea that some meteorites may have come from the Moon is a very old one – more than three centuries ago some scientists suggested that *all* meteorites originated from the Moon. However, samples taken from the Moon during the Apollo missions did not look at all like the meteorites that were then known on Earth and so the idea became lost. Then in 1981 an unusual meteorite was found by an American expedition in Antarctica. ALH 81005 is dark grey, containing centimetre-sized patches of white material. Once the sample was returned to the United States, it was immediately recognised as looking extremely similar to some of the Apollo samples. Since ALH 81005 was discovered, about 20 further lunar meteorites have been found. Most of these were also found in Antarctica but one was found in Australia, and the rest in the Sahara Desert.

The dark and light material present in lunar meteorites correspond to minerals. Looking up to the Moon on a clear night you can see that there are pale-coloured regions, called highlands (made up mostly of a calcium-rich mineral called anorthite), and darker regions, called maria (made of a type of rock called basalt). In lunar meteorites, the dark material is made up of the minerals pyroxene, olivine and plagioclase, and the light-coloured material is mostly anorthite. As well as their appearance, the meteorites were identified as being lunar because the abundances of major and minor elements within them matches both those measured by remote sensing of the Moon, and direct analyses of Apollo samples. The oxygen isotopic composition of lunar meteorites is also identical to Apollo samples. The meteorites also contain abundant nitrogen and noble gases, which were implanted by the solar wind when they resided on the lunar surface. This suggests the meteorites came from a body that has no atmosphere to protect it from the wind blown out by the Sun.

The lunar meteorites may provide much additional information to the lunar return samples. The lunar surface area sampled by all the missions to the Moon only accounts for around 9% of the lunar surface. In contrast, the lunar meteorites come from different regions, and some may have originated on the far side of the Moon, well away from all the previous sample sites. The area visited by the Apollo astronauts may be atypical of the Moon as a whole, and so samples from different lunar regions provide an extremely welcome addition to our knowledge of lunar geology.

Meteorites from Mars

The SNCs are a group of stony meteorites that are thought to come from Mars, rather than the Asteroid Belt. They are all igneous rocks, and are distinguished from other meteorites by their young ages, as low as 165 million years. A young age implies formation on a body that was still active (that is, not totally cooled and solidified) well after the initial aggregation of the Solar System about 4560 million years ago. In other words, the SNCs must originate from a planet-sized body, not an asteroid. There are several reasons why the planet from which they originated is almost certainly Mars (see p. 40).

The mechanism by which the meteorites reach the Earth is by impact ejection: as asteroids impact the Martian surface, craters are formed and if the impactor enters the Martian atmosphere at a sufficiently shallow angle and with a high enough velocity, then ejecta thrown from the surface can escape to orbit the Sun as small bodies in space,

prior to landing on the Earth as meteorites.

The SNCs are named after the three original sub-groups (Shergotty, Nakhla and Chassigny). Collection of additional Martian meteorites from Antarctica and the Sahara Desert has extended the number of sub-groups to five, or possibly six (see p. 42). The different sub-groups are rocks that formed in different locations at or below the Martian surface. The groups have different mineralogies and chemistries and cannot all have come from a single impact event. At least three craters, with minimum diameters of about 12 km, are required to produce the variety of Martian meteorite types.

Shergottites

These are silicate rocks that are now divided into three sub-groups, with different formation localities. One group, the basaltic shergottites, are fine-grained rocks of pyroxene and plagioclase that originated in a lava flow. The other group, the

BELOW LEFT Shargottite Sayh al Uhaymir 008, found in Oman in 1999. The specimen is about 10 cm long.

BELOW Thin section of the Zagami shergottite taken in plane polarized light (field of view is 5 mm). The fractures in the mineral grains (pyroxene) and the paler patches of glass, show that the rock has been shocked.

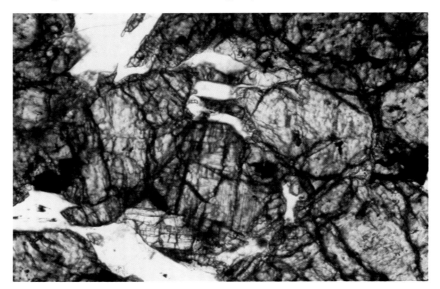

How do we know that the meteorites come from Mars?

The SNC meteorites were recognised as different from asteroidal meteorites long before their Martian origin was accepted. The Martian origin can be deduced through a process of elimination, by considering all the bodies in the Solar System in turn. Several of the planets can be rejected almost immediately. Mercury is too close to the Sun to allow ejecta to escape outwards to Earth. Jupiter, Saturn, Uranus and Neptune are gas planets, and not rocky. The satellites of the giant planets, although rocky in nature, are unlikely source objects since any ejecta blasted from their surfaces will not escape from the gravitational attraction of their parent planet. Pluto is a mixture of rock and ice which, like comets, is not thought to have been molten. So we are left with Venus, Earth and its Moon, and Mars.

BELOW The planet Mars, with the Valles Marineris trough system visible in the foreground.

Venus?

There is small chance of Venus being the parent planet. Venus is about the same size and mass as the Earth and so has a similar escape velocity. But Venus has a very thick atmosphere (about 96 atms, almost 100 times greater than the Earth's atmosphere). In order for ejecta to escape the planet, they must have very high energies when they are ejected from the surface, such that by the time they have traversed the atmosphere the ejecta still have sufficient velocity to escape. Similarly, the incoming impactor must be travelling at great speed so that it is not decelerated to the point where it is unable to impart sufficient energy to the ejecta to enable debris to escape. And on top of all this, once material has escaped from the surface of Venus, it must have the correct orbital vectors to thrust against the gravitational tug of the Sun, and travel outwards to the Earth. Venus, therefore, is dynamically unfavourable as the parent of the SNCs.

Earth?

Impactors frequently hit the Earth, so it could be argued that the SNC meteorites were broken from the Earth's surface with insufficient energy to escape totally. When the rocks fell back to Earth they acquired the fusion crust characteristic of meteorites. The compelling counter-argument to this is that all materials from the Earth have a characteristic composition of oxygen isotopes. A plot of $^{18}O/^{16}O$ against $^{17}O/^{16}O$ shows the oxygen isotopic composition of the SNCs falls on a single line (indicating that they all come from the same planet), but because this line is displaced from the line relating to Earth, the SNCs cannot come from the Earth.

Moon?

The Moon is heavily cratered and none of the dynamic arguments applied to Venus apply to the Moon. But as discussed in the previous chapter, material does come from the Moon and has been identified by its close similarities to Apollo samples. This composition is very different from that of the SNC meteorites. Also, the oxygen isotopic composition in lunar rocks follows the same pattern as that for terrestrial rocks. So the SNC meteorites cannot come from the Moon.

So it must be Mars!

By default, then, the SNC meteorites most probably come from Mars. But there is additional evidence that links them to the planet, which comes from gases trapped within the Elephant Moraine (EET) A79001 meteorite. This meteorite contains inclusions of black glass scattered throughout its mass. The glass was formed by shock melting of mineral grains, presumably during the impact event that catapulted the meteorite from the surface of its parent body. Analysis of gas trapped within the glass during the impact shock shows that it is identical in composition to that of the atmosphere on Mars (as measured by the Viking landers in 1976; see diagram). The only way that this could happen is if EET A79001 came from Mars. Since (on the basis of their oxygen isotopic composition) all the other SNC meteorites come from the same parent as EET A79001, then they too must come from Mars.

BELOW LEFT Diagram showing that the gases trapped inside EET A79001 are the same as, and in the same relative proportions as, those in Mars's atmosphere.

BELOW A cut face of meteorite EET A79001, showing the inclusions of black glass in which pockets of Mars' atmosphere are trapped. The meteorite is about 15 cm across.

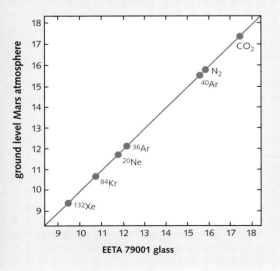

List of Martian Meteorites (April 2002)

Name	Type	Weight (g)	Country	Find/Fall	Date	
ALH 84001	O	1931	Antarctica	Find	1984	
Chassigny	C	4000	France	Fall	1815	
Governador Valadares	N	158	Brazil	Find	1958	
Lafayette	N	800	USA	Find	1931	
Nakhla	N	10,000	Egypt	Fall	1911	
NWA 817	N	104	NW Africa	Find	2000	
Y 000593	N	13,700	Antarctica	Find	2000	⎱ all part of the
Y 000749	N	13,700	Antarctica	Find	2000	⎰ same meteorite
DaG 476	BS	2015	Libya	Find	1998	
DaG 489	BS	2146	Libya	Find	1997	
DaG 670	BS	1619	Libya	Find	1999	all part of the
DaG 735	BS	588	Libya	Find	1996	same meteorite
DaG 876	BS	6	Libya	Find	1998	
Dhofar 019	BS	1056	Oman	Find	2000	
Dhofar 378	BS	15	Oman	Find	2000	
EET A79001	BS	7982	Antarctica	Find	1979	
GRV 9927	BS	10	Antarctica	Find	1999	
LA 001	BS	698	USA	Find	1999	
NWA 1110	BS	118	NW Africa	Find	2001	
NWA 480	BS	28	NW Africa	Find	2000	⎱ all part of the
NWA 856	BS	320	NW Africa	Find	2001	⎰ same meteorite
QUE 94201	BS	12	Antarctica	Find	1994	
SAU 005	BS	1344	Oman	Find	1999	
SAU 008	BS	8579	Oman	Find	1999	
SAU 051	BS	436	Oman	Find	2000	all part of the
SAU 060	BS	42	Oman	Find	2001	same meteorite
SAU 090	BS	95	Oman	Find	2002	
SAU 094	BS	223	Oman	Find	2000	
Shergotty	BS	5000	India	Fall	1865	
Zagami	BS	18,100	Nigeria	Fall	1962	
ALH 77005	LS	482	Antarctica	Find	1977	
LEW 88516	LS	13	Antarctica	Find	1988	
Y 793605	LS	16	Antarctica	Find	1979	
YA 1075	LS	55	Antarctica	Find	2000	
NWA 1068	PS	654	NW Africa	Find	2001	

Key:
BS Basaltic shergottite
PS Picritic shergottite
LS Lherzolitic shergottite
C Chassignite
N Nakhlite
O Orthopyroxenite

lherzolitic shergottites, are more coarse-grained, indicating a slower cooling rate. They formed deeper below the Martian crust than the basaltic shergottites. The shergottites have crystallisation ages of between 165 and 450 million years, indicating that this is when they were erupted on to the planet's surface. At least two cratering events ejected shergottites from Mars into space; these events happened in much more recent times, 0.5–3 million years ago.

Nakhlites

These are almost unshocked rocks that formed at or near the Martian surface in a thick flow or sill. Although they solidified from melts about 1.3 billion years ago, and were ejected from the planet about 10–12 million years ago, the rocks still bear traces of low temperature aqueous processes that can be used to infer conditions on the Martian surface. The meteorites have been altered by weathering, leading to the production of

BELOW LEFT The Nakhla meteorite fell as a shower of stones in Egypt in 1911. Although one of the stones is rumoured to have killed a dog, this story is almost certainly not true.

BELOW Thin section of the Nakhla meteorite (field of view about 3 mm), showing large pyroxene crystals.

RIGHT Thin section of Chassigny in cross-polarized light, showing shocked and deformed olivine grains. Field of view is 3 mm.

secondary minerals (clays, carbonates and sulphates) associated with which are low concentrations of Martian organic material. It has thus been suggested that nakhlites might contain evidence for a Martian biosphere.

Chassigny

This is the only member of its sub-group. It is almost completely composed of the iron and magnesium silicate mineral olivine, and crystallised below the Martian surface about 1.3 billion years ago.

ALH 84001

Like Chassigny, Allan Hills 84001 (ALH 84001) is the only member of its sub-group. It is the oldest of all the Martian meteorites, having crystallised about 4500 million years ago. It has had a long and complex history of shock and thermal metamorphism, and also contains carbonate minerals, indicating that at some stage in its history it has been in contact with Martian water. Since few hydrated minerals (such as are found in clays on Earth) have been identified amongst the alteration products in ALH 84001, it has been proposed that the carbonates were produced at the surface of Mars in a region of restricted water flow, such as an evaporating pool of brine. This hypothesis satisfactorily accounts for the chemical and isotopic characteristics of the carbonates and is also a mechanism that is compatible with an environment in which micro-organisms might survive.

BELOW LEFT A piece of the ALH 84001 orthopyroxenite. It is about 5 cm across.

BELOW RIGHT SEM image (size about 1 μm) of the surface of the orange carbonates in ALH 84001 showing a 'microfossil'.

Are there microfossils in Allan Hills 84001?

Throughout ALH 84001 there are patches of bright orange carbonates, up to a few millimetres across. A team in the USA, led by Dr David McKay from NASA in Houston, Texas, studied ALH 84001. They found tiny structures in the carbonates that looked like fossilised bacteria. Associated with the carbonates were organic compounds. Although some of the organics are probably terrestrial contamination, a proportion are Martian. One particular group of organic compounds, called PAHs, were found in the carbonate in ALH 84001. On Earth, PAHs are produced during the alteration of biological remains in sedimentary rocks. Also present in ALH 84001 are crystals of an iron-rich mineral called magnetite. Magnetite can be produced by bacteria. McKay's team reported that the shapes, with their

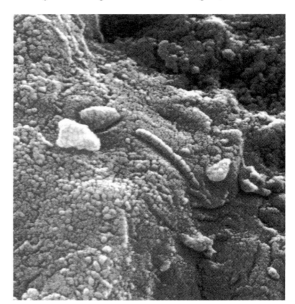

associated organic compounds and magnetite grains, were probably fossilised Martian bacteria. This conclusion has not been widely accepted, as PAHs and magnetite have been found in other (non-Martian) meteorites, where they have non-biological origins. However, the report of possible fossils in a meteorite from Mars has inspired much research, and re-ignited the debate about life on Mars.

LEFT Patches of orange carbonate in ALH 84001. Field of view about 3 mm.

Impact and collision

The amount of extraterrestrial material that falls on the Earth each year has been estimated as between 40,000 and 60,000 tonnes. Almost all of this material arrives as dust grains much less than 1 mm across – about two particles per square kilometre per second hit the Earth. Dust of this size, falling at this rate, poses no hazards at the Earth's surface, although it can cause significant problems in space (see box).

Cosmic dust

The dust grains are called micrometeorites, interplanetary dust particles (IDPs) or cosmic dust depending on how they are collected. Although the typical size of a micrometeorite is less than 30 μm in diameter (sufficiently small to survive passage through the Earth's atmosphere without melting), a large fraction of cosmic dust particles are altered by frictional heating during atmospheric entry. This thermal processing acts to obscure the original nature of the particles.

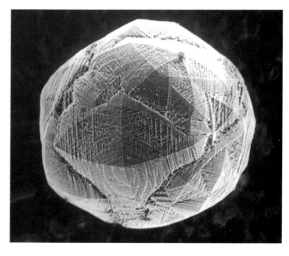

RIGHT A micrometeorite collected from Antarctic ice. Field of view is about 60 μm.

Space debris

Cosmic dust that burns up in the Earth's atmosphere poses no threat. But above the atmosphere, at the height where satellites orbit, even dust particles can be dangerous. Without the atmosphere to slow them down, dust particles travel with a cosmic velocity of about 30 km per second (or 67,500 mph). At that speed, a particle can make a crater in the side of a space vehicle. Degradation of the performance of solar panels can occur and the longer a satellite orbits, the more it becomes peppered with craters from impacting dust particles. There is also a hazard from artificial space debris, in the form of paint flakes and propellant from rockets. Spacecraft such as the *Space Shuttle* and *International Space Station* are constantly at risk from impacting dust particles.

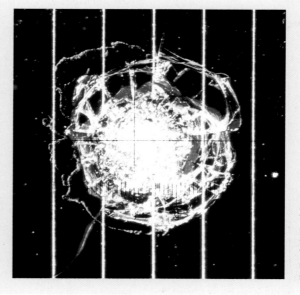

LEFT A magnified image of a crater in a solar cell. The cell is from one of the solar arrays that powered the Hubble Space Telescope, and which was replaced in 1993 during a servicing mission.

The extent of alteration is dependent on the source: dust derived from comets with higher velocities will be more intensely heated than asteroidal dust with lower velocity.

IDPs are collected routinely from the atmosphere by research aircraft operated by NASA flying at altitudes between 18 and 22 km. Micrometeorites may also be collected from several localities on the Earth's surface where the terrestrial dust component is in low abundance. One of the most successful recovery programmes has been through the melting of large volumes of Antarctic ice and subsequent filtering of the water; the Antarctic micrometeorites so recovered are little altered by terrestrial processes.

Cosmic dust represents the sum of material from all bodies within the Solar System (including material from planets, their dust rings and satellites), and also from interstellar dust, although they predominantly arise from collisional debris in the Asteroid Belt, or from dust ejected by comets in their journey past the Sun.

On the basis of telescope observations, some types of IDPs are thought to originate from P- and D-type asteroids, a group of bodies that dominate the outer Asteroid Belt and which rarely give rise to meteorites. Many micrometeorites are also rich in carbon, at levels higher than primitive carbonaceous chondrites. Thus micrometeorites provide a more representative sample of Solar System bodies as a whole.

Intermediate-sized meteorites

Larger bodies impact less frequently – about 1000 meteorites weighing 1–100 kg fall each year. Unfortunately, since much of the Earth's surface is covered by water, or is uninhabited, most of these

TOP An interplanetary dust particle captured in the atmosphere. Field of view is about 25 μm.

ABOVE This young boy was lucky to escape serious injury when the Mbale meteorite fell in 1992.

RIGHT The Peekskill meteorite, and the car which it hit when it fell in October 1992.

meteorites fall unobserved. Only about a dozen are seen to fall each year.

The impact hazard from the fall of a meteorite in the 1–10 kg size ranges is minimal. A person receiving a direct hit would probably die and yet there are no well-documented instances of human fatality from a local meteorite impact. The most widely reported injury associated with impact was that of a young boy in Mbale, Uganda. In August 1992, the boy was playing football with his friends, when a shower of about 50 stones fell. The stones ranged in size from 0.1 g to 27.4 kg, and the boy was hit on the head by one of the smaller stones, weighing approximately 3 g. He escaped with bruises rather than any more serious injury because the fall of the stone had been broken by the leaves of a banana tree.

There have been many instances of buildings and parked cars being hit by meteorites. The

meteorite that fell in Benld, Illinois in 1938 managed to hit both a building and a car, as it fell through the roof of a garage and into the car parked inside. After passing through the floor of the car, the meteorite bounced back and finally embedded itself in the back seat. More recently, in the early evening of 9th October, 1992, a stone meteorite fell onto a Chevrolet parked in Peekskill, New York. The Peekskill meteoroid was accompanied by a bright fireball that was observed and recorded as it travelled across north-eastern United States, before landing in the boot (or trunk) of the car.

As the size of a meteorite increases, the fewer the number that fall each year, but the greater the potential impact hazard. Based on records of meteorites collected over the last 200 years, a meteorite weighing about 100 kg is seen to fall approximately every four years or so; one

ABOVE A stamp issued in 1957 to commemorate the fall of the Sikhote-Alin meteorite in 1947.

BELOW LEFT Part of the Tagish Lake meteorite embedded in ice from the lake.

weighing about 1000 kg (1 tonne), every 50 years or so. Meteorites in this size range are only 50 – 100 cm across and do not form craters.

The biggest meteorite to fall in recent times landed in thick forest in the Sikhote-Alin mountains in Russia, in February 1947. Approximately 23 tonnes of material was recovered at the time and at least a further four tonnes has been unearthed subsequently. The fall was accompanied by a spectacular shower of fireballs, which made over 100 impact holes in the ground, the largest of which was almost 30 m in diameter.

Even more recently, the Tagish Lake meteorite fell in British Columbia, Canada, in January 2000. Only about 1 kg of material was recovered at the time, and a further 10 kg after organised searches several months later. The original entry mass was estimated as 150 tonnes. The low recovery rate for material was a consequence of the meteorite's fragile texture. Tagish Lake landed on a frozen lake and a local hunter collected the first pieces of the meteorite. Shortly after recovery, heavy snowfall in the region prevented further searches, which did not resume until April 2000. It is likely that much material became degraded and simply disappeared during the three months that intervened between fall and recovery.

Craters and crater-forming meteorites

Crater-forming meteorites are those which hit the Earth's surface with sufficient speed (energy) to excavate material. They are the largest meteorites that hit the Earth (objects weighing more than 1000 tonnes and greater than 10 m across). Broadly speaking, meteorites that form craters tend to be iron. They penetrate the atmosphere without breaking into smaller fragments. So, for example, the 50,000-year-old Barringer Crater in Arizona, which is just over 1 km across and about 200 m deep, was made by an iron meteorite approximately 40 m in diameter. An object of this diameter would weigh about 400,000 tonnes. Over 30 tonnes of an iron meteorite, known as Canyon Diablo, have been collected around Barringer Crater, accounting for less than 10% of the mass of the impactor. The remaining material was vaporised on impact, penetrating the surrounding country rock as a vapour of iron. Fortunately, impacts of this magnitude occur only every 100 years or so. But with the spread and speed of growth of population centres, and increasing urbanisation of the landscape,

humanity is more and more likely to suffer a catastrophic cosmic impact.

The mechanism behind crater formation is the reaction of rock to compression and decompression during shock. At impact a shock wave moves forwards into the target rock and also backwards into the impactor, melting and vaporising it. The high shock pressures compress the rocks, rendering them almost fluid. Following the pressure wave is a decompression front, that allows the shocked material to be ejected outwards in a sheet of material. The walls of the newly forming crater are lined with material vaporised from the impactor and the impact site. The floor of the crater consists of broken up and glassy material: rocks that have been distorted or

LEFT Barringer (or Meteor) Crater in Arizona, USA, is approximately 1.2 km across and was produced by an impactor about 40 m wide. The remains of the impactor are known as the Canyon Diablo iron meteorite.

melted during impact. They contain remnants of the impactor.

Simple craters, less than 3–4 km across (of which Barringer Crater is an excellent example), are bowl-shaped, almost always circular depressions and are formed by impactors less than 1 km across. Larger impactors produce more complex craters, several kilometres across. Complex craters may have a central uplift, or plateau, produced as the shocked, unexcavated strata rebound from the impact. Complex craters are sometimes multi-ringed, the rings caused by slumping of the crater walls. The crater floor is often filled with a thick layer of shock-melted glass, or impact melt, which contains trace remnants of the impactor. Some of the most spectacular complex craters, or multi-ringed basins, are seen on the Moon.

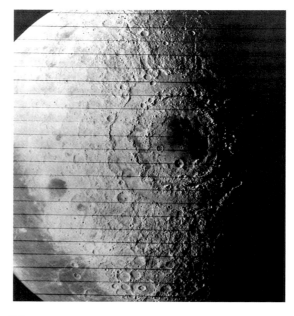

ABOVE LEFT The Mare Orientale basin on the Moon, taken by Luna Orbiter 4. The multi-ringed basin is about 900 km across and was formed by a giant impact early in the Moon's history.

LEFT Aerial view of the New Quebec impact crater with its 250 m deep lake. The crater is about 3.4 km across and formed about 1.4 million years ago.

Shock features

Recognition of craters is often very easy: the bowl-shaped depression of Barringer Crater in Arizona is unmistakable. However, this crater is, geologically speaking, a very young crater in a semi-desert landscape. Craters produced several million years ago might have been buried by sediments, eroded, or otherwise obscured or erased. The shape of the landscape is therefore insufficient to recognise a crater. Fortunately, there are features preserved in impacted rocks that record the impact shock. Shatter cones can be seen in rock exposures that might have been several kilometres away from the impact source. There are also microscopic signs, such as the presence of different forms of a mineral (for example, stishovite and coesite are shock-produced forms of quartz), or deformation features within a mineral.

The occurrence of microtektites (melted beads of glass), and sometimes tiny diamonds is also an indicator of shock processing.

The terrestrial cratering record

Earth has been bombarded throughout its history, but geological and weathering processes have erased or obscured many craters. Currently there are only about 180 well-characterised meteorite craters scattered across the Earth's surface. Most occur on the oldest, continental regions that have been stable for millions of years. The largest craters – at Vredefort in South Africa (300 km in diameter) and Sudbury in Canada (250 km across) – are also the oldest, at about 2000 million years and 1850 million years, respectively. Older craters like these are identified through recognition of shock-produced features in the

BELOW Shatter cones in rocks associated with the Vredefort impact structure in South Africa. The penknife is about 5 cm long.

BELOW RIGHT Magnified view of quartz crystals showing deformation features. Field of view is about 0.5 mm across.

bedrock, since most of the sediments into which the impacts occurred have been eroded away through time. Thus the terrestrial cratering record is incomplete – geological processing and weathering act to obscure and erase craters.

One way of calculating a cratering rate for the Earth is to use the number and ages of craters on the Moon. Since the Moon is so close to the Earth, the cratering rate of Earth and Moon is likely to have been similar. But the Moon is a geologically inactive, airless body (it has no atmosphere to protect it from impactors by slowing them down), and so craters on its surface have been less modified through time than those on the Earth. Based on lunar craters and terrestrial continental

craters, the rate at which craters with diameters greater than 10 km are formed on Earth is between 5–10 craters per million years. The rate is thought to be lower now than in the earliest epoch of Earth's history, when the inner Solar System had not been fully cleared of the debris remaining from planetary formation.

Cretaceous-Tertiary boundary event

Approximately 65 million years ago, at the end of the Cretaceous period, there was a dramatic drop in the numbers of species present on the Earth. This mass extinction has been linked with the collision of a huge meteorite with the Earth. The impact site was at Chicxulub in the Yucatan

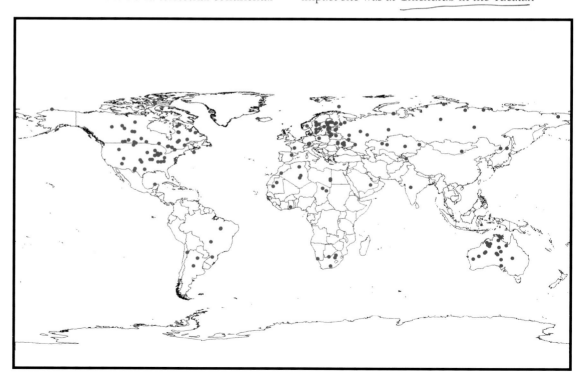

RIGHT Map to show the distribution of major meteorite craters across the Earth's surface.

The Tunguska impact event

Early in the morning of June 30, 1908, a very bright fireball was visible over the remote Tunguska region of Siberia. Contemporary reports record that explosions like thunder were heard. In Russia, Germany and England seismographs recorded tremors and barographs recorded air pressure waves at the time of the explosions. It was not until almost 20 years later, in 1927, that the Russian scientist Kulik led an expedition to the region to search for the remnants of what was presumed to have been a meteorite impact. Kulik did not find a meteorite crater, but an area of devastation about 60 km across. A belt of dead trees that appeared to have been blasted in an intense forest fire surrounded an approximately circular area of swamp. Alignment of the trees, pointing radially outward from the swamp suggested that they had been felled by a shock wave. Only small particles of meteoritic material have been found in peat and tree resin in the area. It is assumed that an impactor (possibly 50 m across) exploded 6–8 km up in the atmosphere and that radiation and a shock wave from the explosion burnt and blew down the trees. It is still not certain whether the impactor was a comet or an asteroid. The energy of the impact has been estimated as 10^{17}J (equivalent to 50 million tonnes of TNT).

LEFT Part of the devastated region of Tunguska following explosion of an asteroid or comet in the atmosphere.

Peninsula in the Gulf of Mexico. The crater is now buried, but geophysical surveys estimate its diameter to be 180–320 km. Environmental effects caused by an impact of these dimensions include massive fires ignited by heat radiated from the impact, a darkening of the sky due to ejected rock dust and smoke, followed by a rapid, global drop in temperature. In the case of Chicxulub, the impact was into sedimentary rock, including sulphate-bearing rocks. This would have resulted in tonnes of sulphur oxides being ejected into the atmosphere. The energy of the impact fused nitrogen and oxygen from the atmosphere into nitrogen oxides. As the temperature dropped, sulphur and nitrogen oxides washed out of the atmosphere as acid rain. These consequences affected the entire globe, not just the local region, and for an extended period of time. It is entirely possible that the global environmental changes caused the extinction of many species, including the dinosaurs, although this theory is by no means completely accepted by many palaeontologists.

Extraterrestrial impacts: Shoemaker-Levy 9

Asteroids and comets bombard all bodies in the Solar System. The presence of an atmosphere slows down incoming impactors and protects a planet, whilst volcanic activity resurfaces a planet. So Mercury and the Moon (inactive and airless bodies) appear much more highly cratered than Mars, Venus and Earth (planets with more recent volcanic activity and thicker atmospheres).

Earth has long been protected against bombardment by the giant planets, whose large gravitational fields serve to scoop up any incoming bodies. Jupiter's role as a defence shield was most graphically illustrated in July, 1994, when comet Shoemaker-Levy 9 (D/1993 F2, or S-L 9) entered the inner part of the Solar System. The comet was in 21 pieces, which travelled towards Jupiter in a regular procession (like beads on a necklace, resulting in S-L 9 being dubbed the 'string of pearls' comet), impacting the planet one after another over several days. The impact speed of each fragment was about 60 km per second (135,000 mph). The explosive energy of the largest fragment (fragment G, the eighth one to hit Jupiter) was approximately 25,000 megatonnes and projected a plume of gas 3000 km up from the planet. The scar left on the face of Jupiter was about 80% of the diameter of the Earth. If any one of the fragments had hit the Earth it would probably have been sufficient to wipe out humanity, as well as a fair percentage of all other land-dwelling species. And so the Earth has been, and still is, shielded from the worst excesses of asteroidal and cometary bombardment by the protective influence of its giant neighbours.

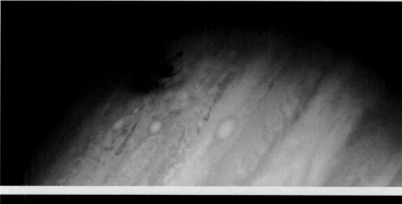

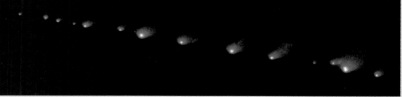

LEFT The scar left behind after fragments of comet Shoemaker-Levy 9 crashed into Jupiter. The shadow is almost the same size as the Earth.

BELOW LEFT Comet Shoemaker-Levy 9 on its approach towards Jupiter in 1994. The fragments are strung out across a distance of about 1.1 million km.

Near Earth objects (NEOs)

At the turn of the 20th century, fewer than 500 asteroids were known. A hundred years later, that figure is over 300,000. With the knowledge of all these orbits came the realisation that there were significant numbers of asteroids that were not confined to the Asteroid Belt between Mars and Jupiter. These asteroids travelled on orbits that brought them close to the Earth. These groups, the Amor, Aten and Apollo asteroids, are classed as the Near Earth Asteroids (NEAs); they currently number around 1500. However, there maybe up to a million or so NEOs (Near Earth Objects) with diameters greater than 50 m.

The formation of craters by meteorite impacts was first proposed in the 1930s, although general acceptance of this hypothesis did not occur until the 1960s. And it is only since the early 1980s that there has been increased recognition of the effect that extraterrestrial impacts have on the Earth and its environment. The perceived hazard from impact of an asteroid or a comet has received increasing publicity over the past decade or so.

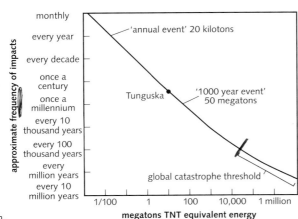

RIGHT Diagram to show the frequency with which high energy impacts occur on earth.

Tektites

Tektites are terrestrial objects that have been produced in an impact event. They are centimetre-sized pieces of natural glass formed by atmospheric quenching of impact melts. Tektites are generally green, brown or black in colour, and have a variety of shapes, most notably like dumb-bells or buttons. The composition of tektite glass reflects the composition of the bedrock into which the associated impact crater was made. Minor trace remnants (< 0.1%) that match the bedrock are also present in the glass. Tektites differ from impactites, or impact glasses, in that the latter are found inside, or very close to the source crater, whilst tektites are produced some distance away from the source.

The area over which tektites are spread is the strewnfield. There are four tektite strewnfields and three of these have been linked to specific craters. The largest strewnfield is that of the Australasian tektites, or Australites; the parent crater of the Australites has not yet been identified.

Description of the global effects of an impact by a kilometre-sized asteroid made it clear that asteroid impact could be a potent force for destruction of life on Earth. Calculations for the potential climatic change that could occur following a nuclear holocaust (the so-called 'nuclear winter') led to the realisation that asteroid impact could cause a greater devastation than the world's combined nuclear arsenal. Following swiftly came recognition of the unpredictability of meteorite impact – it is known that large impactors (greater than 1 km across) strike the Earth, on average, every 50 million years, but that the precise timing of such an event is uncertain. The devastation recorded 65 million years ago, at the end of the Cretaceous period, was probably not a unique event in Earth's history. If that is the case, then an asteroid without warning could hit the Earth at any time. Millions of people would die during such an asteroid strike.

It is now thought that a global catastrophe (which would wipe out humanity) would be triggered by an impact of an asteroid or comet around 1 km across. The challenge is to map the orbits of all NEOs down to diameters of around 1 km. This programme is currently being undertaken by several international groups. One of the problems of tracking an NEO is that several observations must be made before its final orbit is refined. Over the past few years there have been several reported 'near misses'. In order to reduce the danger of crying 'wolf' too frequently, the Torino Scale was devised in 1999. This scale, analogous to the Richter Scale for earthquakes, is used for reporting the potential hazard of NEOs. To date, no hazard greater than 0 on the scale has been reported. The issue of how to deal with a potential asteroid collision is currently exercising the minds of scientists and politicians all over the world.

BELOW Schematic representation of the Torino scale established to report the potential hazard of an impacting NEO.

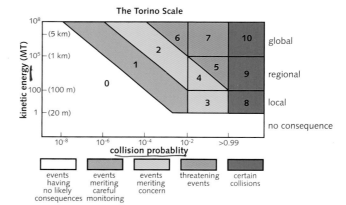

The Torino Scale

Summary and future prospects

Just over 200 years have elapsed since the first recognition that stones fell from the sky. In that time the science of meteoritics has become well-established, and scientists have recognised rocks from the Moon and Mars, as well as the Asteroid Belt, in meteorite collections.

When meteorites were first collected, they were acquired as curios, much as in the way philatelists build up stamp collections. The scientists who studied meteorites did so in order to learn about an individual specimen – and this continues even now, whenever a new meteorite lands. However, meteoritics has developed into much more than the study of discrete meteorites. Now, meteoriticists study a specimen in the wider context of Solar System evolution and development. Throughout this book, we have tried to show how information derived from meteorites can be employed to construct models of Solar System formation. We have used fragments separated from unmelted meteorites (chondrules, CAIs) to infer the type of material from which the Solar System aggregated, and the chronology of events taking place during the earliest stages of Solar System history. In contrast, melted meteorites help illuminate the processes of planetary differentiation and core formation. Lunar meteorites have extended the diversity of moon rocks available for study, and appear to sample regions of the Moon not visited by astronauts or remote probes. Martian meteorites are, so far, the only samples of Mars available for direct laboratory analysis. Studies of components within Martian meteorites have enabled a picture to be built up of primary and secondary processes on Mars, information that is complementary to, and informs, space missions to Mars. Over and above Solar System processes, meteorites yield information on interstellar and circumstellar material.

So, as we celebrate 200 years of meteorite analysis, where will the study of meteorites next lead us? The new generation of laboratory instrumentation promises to be increasingly sophisticated. The ion microprobe, the instrument on which the first analyses of individual pre-solar grains were made, has been developed to even higher levels of spatial resolution. It is anticipated

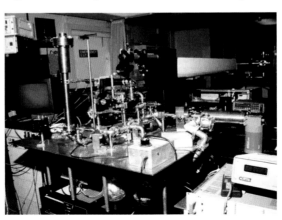

FAR LEFT The nanoSIMS microprobe at Washington University, St. Louis, USA, can measure individual interstellar grains.

LEFT The RELAX mass spectrometer at Manchester University, UK can count individual atoms of xenon.

that the composition of layers and structures within pre-solar grains will soon be measurable. Noble gas studies have played a vital role in untangling the chronology of Solar System events. The development of ever more sensitive instruments, such as the RELAX mass spectrometer at Manchester University, which can count individual atoms of xenon, will allow increasingly precise dating of tiny samples.

But it is not just in the laboratory where advances will be made. Astronomers, astrophysicists and meteoriticists have come to recognise the complementary nature of many of their studies. New ground- and space-based telescopes, with an impressive array of instrumentation, will return data acquired over a range of wavelengths. One example is ESA's Herschel Space Observatory, the largest imaging telescope to be built, that will make observations in the infrared. It will look at dust around newly forming stars, and at processed dust around older stellar objects. Future space exploration missions will return samples from Mars, as well as from asteroids, comets and interplanetary space (e.g. NASA's STARDUST mission). Such materials will fit well into studies based on meteorites, as together all the information is employed to construct more accurate and detailed pictures of how the Solar System formed, and how our star fits into its galactic neighbourhood.

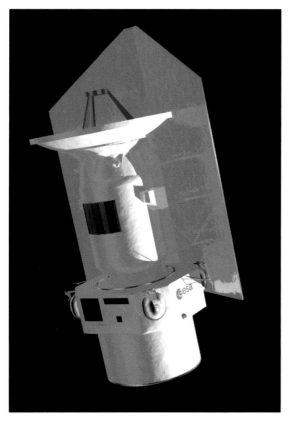

FAR LEFT Artist's impression of the Herschel Space Observatory (launch 2007) which will make observations of stars and dust across a wide range of wavelengths.

LEFT Scientist holding a block of aerogel used in special collectors on the STARDUST space mission. Aerogel is a low density solid material, that will allow capture of dust particles without damaging them.

Further Information

FURTHER READING

Meteorites: A journey through space and time, Alex Bevan and John De Laeter. Smithsonian Institution Press, 2002.

Meteorites and their Parent Planets, 2nd edn., Harry Y. McSween, Jr. Cambridge University Press, 1999.

Meteorites: Their record of Early Solar System Processes, John T. Wasson. W. H. Freeman and Co., New York, 1985.

The New Solar System, J. Kelly Beatty, Carolyn Collins Peterson and Andrew Chaikin (Editors). Cambridge University Press, 1999.

INTERNET ADDRESSES

National and International Organisations
- British National Space Centre (BNSC) http://www.bnsc.gov.uk/
- European Space Agency (ESA) http://www.esa.int/
- National Aeronautical and Space Administration (NASA) http://www.nasa.gov/
- Natural Environmental Research Council (NERC) http://www.nerc.ac.uk/
- Particle Physics & Astronomy Research Council (PPARC) http://www.pparc.ac.uk/
- The Natural History Museum http://www.nhm.ac.uk/

Meteorites and Planetary Sciences research
- Astrobiology web http://www.astrobiology.com/
- Lunar and Planetary Institute http://www.lpi.usra.edu/lpi.html
- Near Earth Objects http://www.nearearthobjects.co.uk

- The Nine planets http://www.seds.org/billa/twn/
- Planetary Sciences Research Discoveries http://www.psrd.hawaii.edu/index.html
- UK Astrobiology Forum http://ast.star.rl.ac.uk/exobiology/

Images
- Anglo Australian Observatory http://www.aao.gov.au/
- Hubble Space Telescope http://hubble.stsci.edu/
- NASA Image Archive http://photojournal.jpl.nasa.gov/
- Nebulae images http://www.seds.org/billa/twn/

Space Missions
Moon
- *Clementine* http://nssdc.gsfc.nasa.gov/planetary/clementine.html
- *Smart-1* http://sci.esa.int/smart-1/

Mars
- *Beagle2* http://www.beagle2.com
- *Mariner, Viking* http://nssdc.gsfc.nasa.gov/planetary/
- *Mars Express* http://sci.esa.int/marsexpress
- *Pathfinder, Global Surveyor, Odyssey* http://marsweb.jpl.nasa.gov/

Asteroids
- *Muses-C* http://www.musesc.isas.ac.jp/INDEX.html
- *NEAR* http://near.jhuapl.edu/

Comets
- *Rosetta* http://sci.esa.int/home/rosetta/index.cfm
- *Stardust* http://stardust.jpl.nasa.gov/

ACKNOWLEDGEMENTS

Robert Hutchison wrote the first edition of this book and is sincerely thanked for his help and support to both of us. We would like to thank Celia Coyne for her editing skills. We are grateful to the Particle Physics and Astronomy Research Council (PPARC) for supporting the research activities of the Meteorites and Micrometeorites Programme.

p.6 bl © Dr Candace Kohl; p.6 br Mike Eaton/© NHM; p.8 t NASA; p.10 © The family of H H Ninninger; p.11 br © Sara Russell; p.12 Mike Eaton/© NHM; p.13 l Courtesy of NASA/JPL/Caltech; p.13 r © Armagh Observatory; p.14 Mike Eaton/© NHM; p.15 l Courtesy of NASA/JPL/Caltech; p.15 r © Prentice-Hall Inc; p.16 © Royal Observatory, Edinburgh/AAO; p.17 t Mike Eaton/© NHM; p.17 b © Bengt Ask; p.18 t from *The Midnight Sky*, Edwin Dunkin, 1872; p.18 b © ESO; p.19 © ESA; p.21 r © Cambridge University Press; p.22 Mike Eaton/© NHM; p.23 l and r D Padgett (IPAC/Caltech), W Brandner (IPAC), K Stapelfeldt (JPL) and NASA; p.25 l and r Chris Burrows (STScI), the WFPC2 Science Team and NASA; p.27 r © Martin Lee; p.28 Rhonda Stroud, US Naval Research Laboratory; p.31 © Birger Schmitz/Mario Tassinari; p.32 Perks Willis Design/© NHM; p.37 l © Caroline Smith; p.40 NASA /Goddard Space Flight Center/NSSDC; p.41 l Mercer Design/ © NHM; p.41 r NASA/JSC; p.44 l NASA/JSC; p.44 r Courtesy of NASA/JPL/Caltech; p.47 t NASA/JSC; p.47 m © Hans Betlem, Dutch Meteor Society; p.47 b © Walt Radomsky; p.48 t issued by The Soviet Union, 20 November 1957; p.48 b Courtesy of University of Western Ontario/University of Calgary; p.49 D Roddy, courtesy of the Lunar and Planetary Institute; p.50 t NASA/Goddard Space Flight Center/NSSDC; p.50 b George Burnside, courtesy of the Lunar and Planetary Institute; p.51 r © Dr Rob Hough, Western Australian Museum; p.54 t and b NASA/STScI; p.55 b David Morrison, NASA; p.56 © Prof Richard P Binzel, MIT; p.57 left © F J Stadermann; p.57 r © Manchester University; p.58 l © ESA; p. 58 r Courtesy of NASA/JPL/Caltech

Title page, back cover and all other images are copyright of The Natural History Museum, London.

AAO, Anglo-Australian Observatory; ESO, European Southern Observatory; ESA, European Space Agency; NSSDC, National Space Science Data Center; STScI, Space Telescope Science Institute.

Index